Frank Bothmann

Rudolf Kerndlmaier

Albert Koffeman

Klaus Mandel

Sarah Wallbank

A Guidebook for Riverside Regeneration

Artery - Transforming Riversides for the Future

This project has received
European Regional
Development Funding
through the INTERREG III B
Community Initiative

ΠWE ENO

INTERREG IIIB
NORTH WEST EUROPE

This publication has received
European Regional Development
Funding through the INTERREG IIIB
Community Initiative

Frank Bothmann
Rudolf Kerndlmaier
Albert Koffeman
Klaus Mandel
Sarah Wallbank

A Guidebook for Riverside Regeneration

Artery - Transforming Riversides
for the Future

Frank Bothmann
Regionalverband Ruhr, Ref. 11
Kronprinzenstrasse 35
45128 Essen
Germany

Rudolf Kerndlmaier
Verband Region Stuttgart
Kronenstrasse 25
70174 Stuttrgart
Germany

Albert I. Koffeman
Provincie Zuid-Holland
Postbus 90602
2509 LP Den Haag
The Netherlands

Klaus Mandel
Verband Region-Rhein-Neckar
P7 20-21
68161 Mannheim
Germany

Sarah Wallbank
Mersey Basin Campaign
Fourways House
57 Hilton Street
Manchester M1 2EJ
United Kingdom

Library of Congress Control Number: 2006932043

ISBN-10   3-540-36725-x Springer Berlin Heidelberg New York
ISBN-13   978-3-540-36725-3 Springer Berlin Heidelberg New York

Springer is a part of Springer Science+Business Media
springer.com
© Springer-Verlag Berlin Heidelberg 2006

Project Coordination, Final Editing, Layout and Graphics: Die PR-BERATER GmbH, Köln
Maps: GeoKarta GbR, Stuttgart

Printed on acid-free paper 30/3141/as 5 4 3 2 1 0

Table of Content

I welcome the opportunity to introduce this guidebook which presents the final results and outputs of the Artery partnership, funded under the INTERREG IIIB North West Europe Programme and which focuses on riverside regeneration across Germany, the Netherlands and the UK.

As this guidebook demonstrates, riversides were for several centuries the backbone of economic development. The Artery partnership brings together five metropolitan regions, namely the Ruhr District, Rhine-Neckar and Stuttgart-Neckar in Germany, Mersey Basin in England's North West and Hollandsche IJssel in the Netherlands. Each represents declining industrial areas which suffer from heavy industrial heritage that hinders re-vitalization. With the decline of major industries such as steel, coal or shipbuilding, their riversides had fallen into neglect and become often inaccessible.

The Artery project partnership demonstrates that this is not an irreversible trend.

As metropolitan regions develop, the need to contain urban sprawl and regenerate riverside regions has made itself more acute. Riversides have increasingly become the focus of regeneration schemes: waterfront locations offer prime locations for businesses, housing, green open spaces for recreational activities. The ten pilot projects within the lifespan of the project have brought tangible improvements to the quality of people's lives and triggered further leverage for investment: in total over 13 million euros. Artery has also developed a number of interesting experiences to develop a sense of ownership and civic pride through community involvement in riverside areas. This guidebook also provides planners and policy-makers with some valuable insights on how to make wider use of Public and Private Partnerships.

Better knowledge of best practice is an essential tool for intelligent urban development. By sharing their findings and experience, the Artery partners have developed extensive expertise on the subject of riverside regeneration and set new benchmarks. They show the way forward for the next programming period in promoting the attractiveness of European cities and regions and achieving sustainable development.

I hope this guidebook will inspire regional developers and policy-makers with fresh ideas and inspiration on how to successfully transform regional riversides for a sustainable future.

I would also like to express my thanks to all the partners engaged in this project for their commitment, their efforts and their spirit of cooperation as well as the programme authorities for assisting them in putting their ideas into practice.

Danuta Hübner
EU-Commissioner for Regional Policy

Text: Frank Bothmann

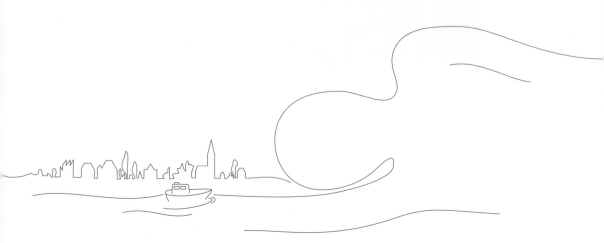

Artery – The Idea

Historical industrial sites from the 19th Century can be found all along the river Ruhr, like the still operating power station in Witten // The Artery projects want to leave a long lasting effect, not only in stone

The Importance of Riversides in Europe

Rivers are Europe's arteries. They are vital habitats for both people and wildlife and have contributed to mankind's development in many different ways. They are connected to human civilisation like no other natural element.

Most early settlements were established close to riverbanks. Rivers provided man with life giving water and a plentiful supply of nourishing fresh fish. They formed the basis for agriculture, providing farmers with water for their fields and farm animals. However, even though settlers were dependent on the river, they kept a respectful distance. On the one hand, this was a natural precaution against high water, but it was also due to the fact that in those days unsecured streams often changed course, meandering through the pristine countryside.

For centuries Europe's inland waters provided food and drinking water, and they were also a means of transport and energy generation. The earliest forms of transport were rafts and boats, which had to be dragged upstream by horse or man power, but nevertheless comfortably transported goods, when roads were unavailable or in bad condition. This was especially practical, if the goods were being transported to or from a mill. Mills housed the first machinery to generate and use water power. This secondary river utilisation was greatly increased during industrialisation.

The Industrial Revolution first took hold of the British Isles in the late 18th Century, spreading into other European countries in the following decades. New technological developments led to highly effective methods of production, and increased levels of performance in all economic fields. However, industries, be it coal, steel or chemical, are dependent on several factors. On the one hand, they need capital investments, local infrastructures and a large market; on the other they need raw materials and a suitable work force. These factors had numerous effects on industrial locations.

Considering all this, rivers were prime locations for industrial sites. For the manufacturing of most products water was essential. It was used for example, to generate power to work machinery, but also as a coolant in steel production. Rivers were also used to dispose of industrial waste, which fed straight into streams and caused heavy pollution. Additionally huge amounts of water were needed for domestic supplies and sewage treatment from urban households.

The expanding trade nations of the 17th Century and the growing industrial regions in the 18th and 19th Century had a great need for efficient means of transport. Before the extension of railways and with roads often being in terrible condition, waterways were the obvious choice. The invention of steamboats made the use of rivers as a traffic route far more practical. In many places the water was not deep enough or bends were too narrow for steamers. In order to make them navigable many European rivers were canalised and straightened. In some cases their riverbeds were even moved.

In the process of industrialisation there was an ongoing structural change from an agricultural society to one focused on mining and industrial economies, as well as the building and construction trade. The industrial economy attracted thousands of labourers and their families. People migrated from the countryside into the growing towns and cities, where industrialisation promised a more dignified life. The process of urbanisation had begun.

As space for labour housing in the town centres close to the industrial plants became scarce, building associations had to resort to building on the riverbanks. Therefore, the respectful distance to the rivers was soon a thing of the past. To avoid flooding, the rivers had to be stopped from meandering. The municipalities started to reinforce the riverbanks to force the stream into a steady flow. It was then possible to build much closer to the embankments.

In many regions urbanisation advanced so far that the waterfronts were no longer accessible. Roads, railroads and houses encroached on the rivers cutting people off. As a consequence opportunities for riverside recreation diminished. Apart from their nutritional and economical functions, this was particularly unfortunate as riversides also have a high social value. Riversides have always attracted people for recreation and leisure activities. Due to urbanisation, many of the places where children used to enjoy the flowing water were demolished. The social culture of promenading along the riverbanks on a Sunday afternoon became lost, as the river literally moved out of peoples' sight.

Centuries of industrialisation have left their mark on urban river landscapes. The effects of river control, pollution and urbanisation are still very visible, even today.

Brownfields are a common sight in many postindustrial regions, like in this abandoned coal-mine in the Ruhr Valley

The Post-Industrial State
of European River Landscapes

Due to technological progress and shifts of energy resources (from coal to oil and natural gas) during the second half of the 20th Century, a dramatic structural change took place in many European countries, affecting both regional economies and industries. In many areas a process of deindustrialisation set in. In this post-industrial era, many regions turned more towards information or service societies.

In this process many of the former uses for rivers and the adjoining land were lost. The transport of goods was largely shifted from water to rail and roads. Various branches of industry collapsed or declined. The formerly heavily used embankments fell into a state of neglect; whole industrial areas lay idle, falling prey to decay. Since then many riversides have turned into rubbish dumps or have become wildly overgrown and inaccessible.

As there were no plans for the abandoned industrial sites and riverbanks, people and businesses turned away from the city centres and moved to the outskirts.

Whole city districts decayed and died away. As a consequence people built new cities or districts at the expense of scarce and valuable landscapes, instead of re-using the former industrial sites.

In order to stop or reverse the exodus of cities or even whole regions and the establishment of new conurbations elsewhere, regional development plans, sensitive to the economical structure of the region, have to be elaborated and implemented.

In the past decade awareness of this deplorable situation has greatly increased. The need and demand to change conditions has been thoroughly discussed. More and more organisations and authorities have come to the conclusion that riverside transformation is an important issue and should be at the heart of regional strategies in the years to come.

Today there are still various industrial logistic sites operating along the Neckar

Water Framework Directive

~ Overarching system with its main concern to protect all European waters

~ Sets clear objective that a 'good status' for all European waters must be achieved by 2015

~ Helps to secure European water resources for today and for future generations

~ Focuses on sustainable water use throughout Europe

~ Legislation developed by all water concerned parties, from agriculture and industry to environmental organisations as well as local and national authorities

~ Acknowledges the need to involve the public in order to secure the enforceability of objectives

The Importance of Riverside Regeneration in Europe

For various different reasons the regeneration of European river landscapes is a very important issue at the turn of the 21st Century. The necessity for transforming riversides results from social, economical and environmental effects.

Regional and economic planning and development have had to take into account that derelict unattractive sites and inaccessible riversides do not promote a favourable image of the region or invite new businesses and people to settle. Also the local population will start looking for alternatives and leave the region or at least move to the outskirts.

Riverside development is also very important on a social dimension. Since the beginning of the industrialisation process, about 200 years ago, the per capita income in North West European countries increased approximately twelve-fold. Aligned with improved income were people's living standards and educational background. Additionally due to automatisation processes, average weekly working hours decreased. The majority of people had more time and money to spend on leisure. These circumstances contributed to an increased appreciation of their cultural heritage and natural surroundings. However, at the same time nature as a recreational resource had become a scare product. Thereby derelict, neglected and underused riversides came more into the focus of regional and spatial development strategies. By transforming former river-based industrial sites and making riverbanks accessible to people, cities and conurbations improve their attractiveness and offer locals inviting places to retreat to.

The impoverishment of the natural diversity is another important key factor. The centuries of long ongoing destruction and pollution of natural habitats has forced much of the local wildlife to retreat. Such environmental consequences can be stopped or partially reversed through actions aimed at re-naturalising the riverbanks and enhancing the water quality of the rivers.

Water quality and water resource availability are essential issues affecting both people and wildlife alike. Only 2.5 percent of the world's water is fresh water and only 0.3 percent can be found in surface waters like rivers, lakes and reservoirs. The remainder is locked up in groundwater, ice sheets and glaciers. It is therefore important that this finite resource is neither wasted nor polluted.

However, due to urban and industrial developments, the state of European waters is far from satisfactory. According to studies by the European Commission 20 percent of all inland waters are still seriously polluted in the early years of the 21st Century. Consequentially we are not only poisoning fauna and flora but eventually ourselves.

From the 1970s onwards a growing ecological consciousness made people more aware of their environment being a finite source. The scarce commodity of green spaces in densely populated industrial regions made many residents long for local recreational areas in the form of 'untouched' nature and rivers.

Fly-tipping still causes severe pollution along the Mersey banks

Artery – The Idea

The special role river landscapes hold for people and wildlife has been acknowledged by authorities as high as the European Union. The protection and redevelopment of inland surface waters has been included in several decrees. Most important among them are the Water Framework Directive (WFD), the Fauna, Flora and Habitat (FFH) and the European Spatial Development Perspective (ESDP). It is now up to the people, local and national authorities and NGOs to translate these plans into action.

Artery: Joint Forces for Maximum Effect

Artery: Ten projects, five regions, four rivers, three nations and one mission. In 2003 sixteen partners joined forces to set a new benchmark in riverside regeneration. Over a period of three years, with a budget of 12.5 million euros they set out to transform river landscapes across the five regions Ruhr Valley, Rhine-Neckar and Stuttgart-Neckar in Germany, the Mersey Basin in England's Northwest and the Hollandsche IJssel in the Netherlands.

These regions have all suffered from the consequences of (post-) industrialisation. Although there are a diverse number of reasons for riverside regeneration the Artery partners share common problems. One major issue that concerned all five regions was the accessibility of the rivers. In the Ruhr Valley and the Mersey Basin economic structural changes caused problems on a social, economic and environmental level. The decline of the traditional industrial sectors – in the coal and steel industries in the Ruhr-area; in transport, distribution and shipping in the Mersey Basin – triggered a high loss of employment. The retreating industries left environmentally damaged land and derelict sites that lay idle for a long time. Creating clean green spaces as well as new recreational facilities was at the heart of these two regions.

The main reasons for riverside regeneration along the Hollandsche IJssel were like in England, social and environmental problems. For both it was important to increase the accessibility and spatial quality of former industrial rivers on the one hand, and to start a large cleanup on the other. While the water of the river Mersey was highly polluted, in Holland it was the soil of the riverbed as well as along the embankment that was seriously contaminated.

In the Rhine-Neckar region from the cities of Mannheim to Heidelberg it is the aim to start a sustainable river landscape development along the Neckar and to promote greater opportunities for leisure activities for the residents. The Stuttgart-Neckar region placed ecological re-naturalisation in the foreground.

Another problem the five partner regions are dealing with is a population that is hardly aware that it lives next to a river that was once the artery of its region. With Artery the local initiatives strive to increase their regions' success in creating sustainable environments that in turn contribute to building new economic opportunities, green spaces, and leisure opportunities. They are determined to encourage people in municipalities, towns and cities to 'turn back to face the water'.

Working across Europe with Common Strategies

In the five partner regions, local organisations and associations recognised the need to put riverside regeneration at the heart of regional development. Many ambitious projects had been started in numerous places. Different organisations in the different countries had approached the topic with different strategies.

These strategies proved to be successful in their own ways, and to maximise the effect of the efforts, the Artery partners decided to bring their knowledge on a European level together. With the four common strategic approaches; public participation, public-private partnerships, regional strategies and public awareness, they aimed to facilitate the ten pilot projects.

These common themes are the platform for transnational exchange and have encouraged co-operation between the regions. With these strategic approaches Artery wanted to be able to include everybody affected by the actions taken. The strategy 'public participation' aimed to get the local public involved, while 'public-private partnership' approached the local authorities as well as the regional private sector. This strategic methodology will ensure the sustainability of the Artery pilot projects.

This model of working together with four common strategic approaches underlines the nature of the Artery programme. The partner regions understood

Beforehand the straightened Neckar in Mannheim only offered a few attractive recreational opportunities for residents

themselves to be in a 'learning partnership'. Four of the five partner regions were a thematic leadpartner in one of the common themes. In this field they were the 'experts'. In the other strategies they saw themselves as 'learners', profiting from the knowledge of the expert regions. Through workshops, reports and personal exchange the partners engaged in a constant interaction. The common approaches facilitated learning across Europe. The interactions brought forward a greater transnational understanding and cross-regional network and have resulted in the successful realisation of ten truly European pilot projects.

Future regional development projects in other European regions can benefit from the knowledge and experience gained in the Artery project. It is Artery's overarching goal to realise regional projects in an innovative way and thereby enhance the status of their locations, and also to inspire other local initiatives to follow the same course. It wants to support new, stronger regional and cross-regional partnerships and networks.

With the successful completion of the ten pilot projects through the common strategies approach, Artery shows that riverside regeneration pays off, making central urban areas available once again for recreational and leisure opportunities as well as restoring local wildlife habitats. Riverside leisure and recreation in return bring jobs, health and vitality to the municipalities and the regions as a whole.

Artery – A European Project

As Artery brings together regional initiatives concerned with river landscape development, the transnational aspect is very important. Beside establishing long lasting transnational contacts and sharing experiences, Artery's work contributes to the new planning culture requested by the ESDP issued by the European Commission.

The European Commission values transnational co-operation. As in the case of Artery the exchange and common elaboration of successful working methods furnishes concrete results. It increases the value of the partner regions and thereby contributes to the worth of the whole European Community.

A joint venture like Artery is an effective way for local initiatives to develop their own region. It also enables them to participate in the construction of Europe

What is INTERREG IIIB North West Europe (NWE)?

~ An EU community initiative programme that ran between 2000 and 2006

~ Furthers transnational co-operation between the countries of North West Europe with European Regional Development Fund (ERDF) co-financing

~ Supports projects that strive for sustainable developments in the North West of Europe

~ Funds for various areas of transnational co-operation in regional development, such as urban development, transportation, protection and sustainability of natural resources and cultural heritage, as well as promoting European ports and seas

~ NWE has a budget of 330 million euros

through favouring exchanges, working out innovative methods and solutions to problems that concern a group of regions in North West Europe that share more than just a geographic area.

The European INTERREG IIIB North West Europe (NWE), part of the INTERREG funding programme, acknowledged the innovative character of Artery and its contribution in constructing Europe. Therefore, it rewarded this transnational joint venture with a total of 5.4 million euros.

Artery: The Organisation

For regional planners and local initiatives the question arises, how can a joint venture work with so many

The European project partners in Artery met regularly to share their experiences, like on the river Ruhr

parties involved? How can a co-operation of partner regions that are so far apart succeed? How can finances be managed or the transfer of knowledge be guaranteed with lingual and cultural differences involved?

In order to manage a diverse and trans-regional project like Artery, a clear structure is needed, as well as someone who co-ordinates the work of the regional partners. Clear rules of communication have to be established at the start. Everybody needs to have a clear understanding of his duties, be it elaborating workshops or preparing papers or minutes. Tasks as well as responsibilities have to be organised, in order to avoid important issues being forgotten or resources wasted if various parties do the same job.

Artery's organisational structure provides just such a clear structure and will at this point be briefly introduced to clarify the terms and specifications used in the following chapters.

The thematic leadpartner is responsible for elaborating and documenting the one common strategic approach of Artery to which his or her region is 'expert'.

The regional leadpartner represents the region on EU level. Artery being a learning partnership, the five partner regions are all on the same level of hierarchy.

The Artery pilot projects are realised by local initiatives, organisations and regional authorities; the pilot project partners. As there are several pilots implemented in one region the regional leadpartner is in charge to co-ordinate among and report to EU level.

The Regionalverband Ruhr (Regional Association Ruhr), being the regional leadpartner for the Ruhr Valley, is at the same time overall project leadpartner and co-ordinates the work of the regional and thematic leadpartners as well as reporting to the INTERREG IIIB NWE Programme Secretariat in Lille, France. The leadpartner is supported on an operational level by the Project Group where the regional partners meet regularly and secure the smooth running of the project's implementation.

The Artery Advisory Board was set up to ensure direct links to the respective regional political bodies.

1949 river bath in Wetter; river bath culture has a long tradition at the Ruhr

The board consists of one political representative from each respective partner region. Its main function is to ensure the region's political backup and give advice on the projects development which has proven to be very helpful.

Managing a 12.5 million euros budget makes sound financial management an absolute necessity. Financial controlling instruments and a clear reporting procedure have proven to be essential for successful financial management.

The Ruhrtal

The Ruhrtal (Ruhr Valley) is the heart of the Ruhrgebiet (Ruhr area) industrial area with roughly 2.5 million people living in the area of the densely populated valley. The river Ruhr is the artery of the region. Not only does it provide more than 75 percent of the region's drinking water, but without it the establishment of the coal and steel industries would not have been possible.

The development of the coal industry began in the Middle Ages and was greatly expanded with the invention of the steam engine in the middle of the 19th Century. For the following 150 years mining, coal and steel industries defined the development of the Ruhr Valley.

Everything in the region was focused on the industrial sector. The regulation of the water through canalisation, reservoirs and weirs made it possible to build industrial plants and roads on the flood plain. Even today power plants, supply and disposal facilities as well as energy lines dominate the former pastoral landscape. In other places the river was canalised and the riverbed moved to make it navigable. Additionally a railroad network along the river was established to guarantee the transport of raw materials as well as processed goods.

After the canalisation of the rivers and the high spatial demands of the expanding steel industry, new industrial areas were developed on the now high water free, secured riverbanks. Today these often post-industrial areas disconnect the cities from 'their' river. Because some of these areas have changed to brownfields there is now the chance to bring new development to the sites and to turn the cities around to face the river once again.

The River Ruhr

~ 214 km long right-bank tributary of the Rhine, catchment area of 4,434 km²

~ Rises in the Sauerland region and flows into the Rhine in Duisburg.

~ With five reservoirs and many weirs, it is the most important drinking water supply for the whole Ruhr area.

~ Navigable via floodgates, for motor ships only from the estuary up to the Baldeneysee in Essen.

~ Since the 1970s promoted as a recreation area for the resident population as well as tourists.

~ The Ruhr Valley is a 40 km stretch of the river, developed in the partnership called 'Das Ruhrtal' which includes the six cities Bochum, Hagen, Hattingen, Herdecke, Wetter and Witten.

After the Second World War the Ruhr area was the engine of reconstruction as it provided energy and steel. The region also contributed strongly to the economic miracle in Germany that is also known as the 'Wirtschaftswunder'.

However, in the second half of the 20th Century the coal and steel industry started to decline and left a legacy of devastated land, abandoned production centres and the loss of whole sectors of work, which resulted in great economical and social problems. Until reunification the Ruhr area held the highest unemployment rate in former West Germany.

Since the 1970s, to counteract the effects of deindustrialisation, an economic structural change towards the tertiary sector has been driven forward. Local authorities and the Regionalverband Ruhr (Regional Association Ruhr), or RVR, worked out strategies to

The River Mersey

- Length: 111 km, 26 km of estuary, catchment area including its tributaries 4,680 km^2 with a population of over five million people

- The river Mersey starts at the confluence of the river Tame and Goyt in Stockport, before flowing west towards Liverpool where it opens out to form a wide estuary of more than 75 km^2 to enter the Irish Sea.

- The Mersey is tidal up to Howley Weir at Warrington, and has a tidal range of about 10 m, the second highest in the UK.

- Partially canalised up to Warrington to make it navigable.

- In the 1980s the Mersey Estuary was the most polluted estuary in Europe - it has now been declared as the most improved.

- The Mersey Estuary is a designated RAMSAR site and listed as a Special Protection Area (SPA) because of the range of habitats, especially for wintering birds, wildfowl and waders.

---

enhance the service supply economic sector. In order to keep the population from moving away from the Ruhr area, new economical opportunities like IT, media and logistics as well as science and research, have been supported, and on top of this the Ruhr Valley is also being heavily promoted as an attractive recreation area and tourist destination.

Although a lot has already been achieved there is more work ahead; especially on the shores of the Ruhr. Derelict industrial sites still sully the view and block access to the embankment. Riverside regeneration is therefore an important element of the regional development plans. Taking this challenge into account in 1998 six cities along the river Ruhr established an informal partnership to commonly develop the natural and heritage assets of the valley for leisure and new tourism services.

Thanks to the efforts of the Regional Association Ruhr and financial support from European Structural Funds, the Ruhr Valley as well as the whole Ruhr area is today, once again, a promising European region. Despite the decline of heavy industry the Ruhr area is still the most important German and European industrial district. And even though the process of deindustrialisation is visibly in progress, further struc-

tural changes will improve the image and the quality of life of the region.

The Mersey Basin

The Mersey Basin in England's Northwest is one of the most densely populated urban areas in Britain. It encompasses Liverpool, European Capital of Culture 2008 and Manchester.

The river Mersey not only gave the region its name, but its riversides became a prime location for industrial expansion after the opening of Liverpool's first dock in 1715. With the construction of four more docks in the following years the seaport grew bigger than its London counterpart. In the following century the seaport gained major significance for passengers and freight transportation and turned Liverpool into an important trade centre for the British Empire. Railways and canals, including the Manchester Ship Canal, were constructed to transport goods to and from industrial areas and throughout Great Britain.

In the same century the industrial revolution set in with mechanical spinning and weaving facilities. Mills

---

Tom Workman, president of the Liverpool Sailing Club, inspecting the burnt-out ruins of the former club house, which was destroyed in an arson attack

were established to work the new machines and technologies with water power. These were the early days of the textile industry, which would soon become an important industrial activity in the region. Aligned with this growth came an increase in the bleaching, dying and finishing trades and thereby in the heavy chemical industry. Other industrial pillars in the region were transport, distribution and shipbuilding. All of these industries used the river as a means of transport and waste disposal, even for particularly noxious substances such as mercury.

After the Second World War and with the decline of the British Empire the significance of Liverpool's port decreased and in the following decades various branches of trade and industry disintegrated. Huge wastelands as well as the most severely polluted river in the whole of Europe were left behind. Until today the authorities in the Mersey Basin face the complicated tasks of restructuring the impoverished economy and fighting a high unemployment rate. At the same time the region has had to fight the industrial pollution of the Mersey and its estuary.

It was recognised that environmental assets occupy an important role in the regeneration process and this underpinned the establishment of the Mersey Basin Campaign (MBC). Michael Heseltine, then Environment Minister in Margaret Thatcher's Conservative government, was the driving force behind its creation. At the time, he spoke of the river Mersey as "a disgrace to a civilised society." There was no single driving force behind riverside development; a single catchment wide campaign enabled all interested partners to work towards a common objective.

However, since the 1980s the process of regenerating the Mersey Basin has been underway and a lot has already been achieved. An active partnership, including riverside businesses, regulatory authorities, government agencies and local communities, set out to develop a sustainable economy and the regeneration of England's Northwest.

A major focus has been placed on regenerating abandoned industrial sites to create opportunities for development and community use. These reclaimed sites will attract investment and new businesses and thereby create new jobs for the local population. The reclaimed land also provides new and enhanced facilities for the local communities. Clean-up actions protect environmentally sensitive areas from pollution and help to create a sense of local ownership for the river.

The work of organisations such as the Mersey Basin Campaign (MBC) and the North West Development Agency (NWDA), are supported by local authorities and the European Union. The success of projects like Artery leading to new regeneration and investment in the environment, will contribute once again to the flourishing of England's Northwest.

The Rhine-Neckar Region

The Rhine-Neckar region is an economic metropolitan area in the catchment of the Neckar estuary. It encompasses the cities of Mannheim, Heidelberg and Ludwigshafen. With 2.4 million residents and 5,637 km^2 it is the seventh largest economic area in Germany.

To a large extend the region covers the territory of the former 'Kurpfalz' (Palatinate). Due to these roots close socio-cultural co-operation and regional consciousness are still intact, despite the region's split into three states (Baden-Württemberg, Hessen and Rheinland-Pfalz). These three states signed an inter-state treaty to facilitate co-operation in the region on a social and cultural level, to promote common economic ventures and to implement concerted regional planning. This treaty underlined the acknowledgement of the Rhine-Neckar as a European metropolitan region.

The buzzing Rhine-Neckar region has a balanced system of industrial and service economies. This development can be attributed to its logistically convenient location close to the two main waterways Rhine and Neckar. The Mannheim/Ludwigshafen river port is the second largest inshore harbour in Germany and a trade centre for goods from all over the world.

It wasn't only the service and trade sectors that boomed on the shores of the Neckar and the Rhine. Even in the early days of industrialisation the production industry prospered. Many of today's global players from the chemical and technological industry

River port of Mannheim 1925: The river Neckar and its banks were vital for the development of industries in the region // The Neckar between Seckenheim and Ilvesheim in the 1990's: Here regeneration measures were successfully implemented

have been based in the region for the last 200 years. Further traditional industrial fields that were established over the last century are pharmacy, engineering and electronics, print and food processing. The two streams here played an important role as they guaranteed a sufficient water supply and a means of logistics and waste water disposal.

However, what got lost in the exemplary development of this region were local recreation areas. The urban spatial structure has no large connected green spaces in the centre of the conurbation. For example due to roads and buildings the city centre of Mannheim is severely cut off from the banks of the river Neckar. Steep stone reinforcements at the shoreline hinder walkers from reaching the water. Although there are vast grasslands on the riverside, there are no lawns for sunbathing or playing fields for children. Consequently many people are not even aware of the river on their doorstep.

Riverside regeneration and raising awareness for the river are vital for the attractiveness of the region. They are therefore essential to regional strategies promoting the Rhine-Neckar metropolitan area as a regional competitor within Europe. The Verband Region Rhein-Neckar (Association Region Rhine-Neckar) and the Nachbarschaftsverband Heidelberg-Mannheim (Neighbourhood Association Heidelberg-Mannheim) have already started realising plans to create local recreational areas as well as an adjacent green band along the river. In this inter-communal co-operation of cities and municipalities along the Neckar, various smaller and larger projects have already been realised, many more will follow.

The Stuttgart-Neckar Region

As the name suggests; the Stuttgart-Neckar region centres around the capital of the state Baden-Württemberg and the river Neckar. With 2.6 million people living over 3,654 km² it is one of the most densely populated regions in Germany and Europe.

Its economic system is based on medium-sized companies in the tertiary sector as well as on industry. In the tertiary sector there is a clear focus on finances and public and private service supply. In the industrial sector the car and engineering industries are still dominant.

It might be this mixture of industry and service economy that have kept the region economically healthy while other industrial regions have had to struggle with structural changes, but maybe it can also be put down to the fact that industrialisation set in later and certain branches are still in operation. There are not many declined industrial sites in this urban populated area along the Neckar.

It is therefore not so much the problem of economical structure that local authorities are concerned with, but more a social and environmental one. Like in the Rhine-Neckar region, industrial sites, roads and railroads have been build close to the riverbanks and left it inaccessible to the local population. The Verband Region Stuttgart (Association Region Stuttgart) has therefore developed an overarching regeneration plan – the concept Landscapepark Neckar. The associa-

The River Neckar

~ The 367 km long right-bank tributary of the Rhine has a catchment area of 14,000 km².

~ The river Neckar rises in the Black Forest near Schwenningen and flows into the Rhine at Mannheim.

~ The river is navigable on the canalised stretch of 203 km, from the town Plochingen to the estuary in Mannheim, 27 dams with a total height difference of 160.70 m.

~ The river Neckar is one of the most important German waterways.

~ Its waters are used as process water as well as for generating water power.

~ Severe interferences, like canalisation and straightening, have had a great impact on the natural ecology and aquatic river habitat.

tion is determined to reclaim green spaces along the Neckar and create new recreation areas.

In order to do this, another issue has to be dealt with. The river has been heavily altered over the course of industrialisation. It has been canalised, its riverbed shifted. With the construction of dams the Neckar has been made navigable. These interferences with the natural flow of the water have had numerous consequences. Wildlife habitats were destroyed. As a high water precaution steep bank reinforcements were erected which not only prevented water from coming over the embankments, they also stopped people getting to the river. Through canalisation the water flow increased. This fast flowing stream caused problems for spawning fish and as the water digs deep into the riverbed the eroded sediments are deposited further downstream often blocking harbour entrances.

The regional association has taken actions to give the river back some of its naturalness. Restoring the river's natural habitat not only contributes to a better ecological balance in the region, but is the basis for creating the above mentioned recreation areas.

With these actions the Association Region Stuttgart started to reverse some of the effects caused by human interference. Measures to re-naturalise the river and its embankments contribute to make the region a more attractive place. This will prevent people from leaving the area to retreat to more rural towns, and will also attract people who want to live in a region that has a lot to offer, including green spaces to enjoy nature.

The Hollandsche IJssel Region

As she flows along, the Hollandsche IJssel passes the densely populated Randstad agglomeration as well as the open spaces of the rural Green Heart Region.

The river used to facilitate trade and commerce in the region allowing typical industries to develop. Shipyards and related industries settled on the shores of the Hollandsche IJssel from the 17th Century onwards and the river provided the clay for an exten-

Canalised Neckar at Altbach. Dams are preventing fish population from wandering upstream

The Hollandsche IJssel

~ The tidal part of the river runs from Gouda until it joins the Nieuwe Maas river just before Rotterdam.

~ The Hollandsche IJssel no longer receives water from the Rhine. It draws its water from the drained polders and therefore plays an important role in the household water supply for the region.

~ The Hollandsche IJssel has been fully aligned with dikes for centuries and through the continued draining of the polder land the weak peat soil set and the river bed rose above its surroundings. A result of this is that the lowest point of the Netherlands is located in Nieuwerkerk aan den IJssel, situated at 6.74 meters below NAP, the average sea level.

~ Through its open connection to the North Sea it is tidal, but the silted North Sea water comes no further than Krimpen aan den IJssel. The freshwater tide leads to a difference in water levels of approx. 2 m.

was long characterised by its sparse population and agriculture due to its isolated position between the branches of the Rhine estuary. With the opening of the bridge, Krimpen aan den IJssel became easily accessible from Rotterdam and attracted new residents. Housing and recreational needs for the commuters of the Randstad now intensified pressure for open spaces to be developed.

The towns along the north shore are very well connected with Rotterdam and Gouda by motorway and rail and therefore saw a different development in the 20th Century. Where for centuries the river was the backbone of infrastructure, companies now preferred locations along the motorway or left the traffic congested area altogether to settle in the port of Rotterdam. Linked to the Rotterdam metro system, Capelle aan den IJssel turned into a commuter town, blurring the boundaries of its large neighbour.

Of special concern along the Hollandsche IJssel was the contaminated soil. After World War II the need for suitable ground for development led to the reclamation of land along the river. It was cheaper to heighten the tidal riverbanks, the 'zellingen', than to stabilize the weak peat soil of the polder. After the decline of traditional industries the precious land along the river banks was allocated for new housing areas. But the material used here to heighten the riverbanks turned out to be heavily polluted, indicating the general neglect of the region and its river. The soil was in urgent need of sanitation, as Holland's most polluted river called for action.

The Projectteam Hollandsche IJssel was initiated to tackle the challenge. The sustainable development of the region is now based on four main functions: living, working, recreation and nature. The decontamination and regeneration of the river and its banks is altering the perception of the Hollandsche IJssel. Motivating citizens and companies to take a stake in the region, the initiative strengthens the regions qualities and turns the long neglected river into the artery of the region once again.

sive brick industry from its own riverbed. This explains why the names of most villages in the region end with 'on the IJssel'.

It is interesting to note that there is not much of a notion of a Hollandsche IJssel region, neither on an administrative level nor amongst the public. Though an important artery, connecting communities along its shores with trade partners in Holland and along the Rhine, the Hollandsche IJssel has always been seen as a barrier rather than a binding element. Up until 1958, when the Algera Bridge between Capelle and Krimpen opened, the only way to cross the river was by ferry or a bridge 20 kilometres to the north at Gouda.

Accordingly the opposing shores underwent notably different developments. The polder on the south shore

Vuyk shipyard 1940: For generations the shipyard dominated the river front in Capelle aan den IJssel // On the verge of finishing the new community park on the site of the former shipyard

Artery – The Projects

GB //
Speke and Garston
Coastal Reserve

Artery turns former airfield into a new recreational
resource for local people and improved habitat for
wildlife. The new state-of-the-art clubhouse transforms
Liverpool Sailing Club into a community based premier
water sports facility.

The new club-house of the Liverpool Sailing Club is shaped like the
sails of a sailing boat // In dialogue with the local youth to gain support
for the redevelopment of the Mersey Estuary // Old airfield makes room
for nature and recreation

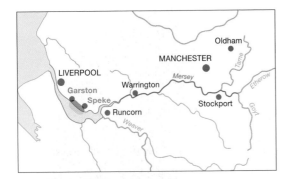

Great Britain
Northwest England
Speke and Garston
River Mersey

From the edge of the bustling new Liverpool John Lennon Airport, an urban wasteland used to stretch across the site of the city's previous airport down to the banks of the famous river Mersey. Burnt out cars littered the view of the ecologically sensitive Mersey Estuary. Illegal fly-tippers had dumped scrap and rubbish all along the river. Belligerent youngsters roamed the terrain.

A safe area for both people and wildlife – that was what the Mersey Basin Campaign (MBC) and landowner Peel Holdings had in mind for the derelict land next to the stunning Mersey Estuary, which is an internationally important Special Protection Area for migratory birds. "The site is of high ecological importance. It was clear that we needed to create a safe environment for people and a good natural environment for wildlife", states Mersey Basin Campaign development manager Iain Taylor.

"Not long ago the land, which is now the Speke and Garston Coastal Reserve, used to be a 'no-go' area", explains Louise Morrissey, Peel Holdings' head of Land and Planning. "After the airport was relocated, the land fell into decline. Fly-tippers moved in and a lot of anti-social issues went on." The Liverpool Sailing Club, which has been on the site for almost fifty years, was regularly vandalised and in the year 2000 was finally destroyed in an arson attack.

In order to finance the one million pound project the Mersey Basin Campaign enlisted the help of public-private partnerships. Landowner Peel Holdings not only offered their land but also legal advice and management. For the coastal reserve's maintenance a special company has been established to guar-

antee that the improvements are not short lived, the Speke and Garston Coastal Reserve Management Company. This unique organisation aims to involve local companies in the regeneration of the area and long term management.

Cleaning up the embankment and re-naturalising the land was not only important for the wildlife, but also for Speke and Garston. As two of the most deprived communities in England, they benefit from the new coastal reserve in more ways than one. The now secure area offers people a green space and a unique access point to the river Mersey. The safe environment has also helped to increase the growth of the neighbouring business park.

The new Sailing Club clubhouse, also part of the Artery programme, is open to everyone and offers the opportunity to learn about sailing and other water sports. "Here in this very natural vicinity many people, especially the young, can discover new ways of experiencing their surrounding environment", explains Tom Workman, president of the Liverpool Sailing Club. Even before the new club building opened, land-based sailing skill activities like blokarts (small land yachts) were up and running. These facilities are accessible for children and adaptable for disabled people.

In close partnership with Peel Holdings, Mersey Waterfront, Liverpool City Council, Northwest Development Agency, and Artery, the Mersey Basin Campaign has succeeded in transforming a vast wasteland into a flourishing riverside, and a burnt out shell into a local meeting point for both young and old.

GB //
Mersey Vale Nature Park

The busy M60 motorway through Stockport used to offer drivers a bleak vista of derelict industrial sites clustered along the banks of the river Mersey. The view today is very different - a green waterside nature park offering a haven for wildlife and a place for local people to enjoy.

Land of former railway sidings reconnects now the community of Heaton Mersey with the river // Re-naturalised site becomes part of the nationwide Trans Pennine Trail

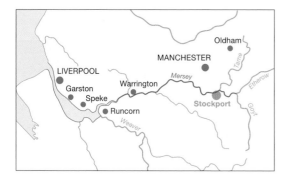

Great Britain
Northwest England
Stockport
River Mersey

The River Mersey was once heavily industrialised along much of its length and the Mersey Vale area of Stockport was no exception. The crumbling abutment from a derelict railway bridge offers a reminder of the area's history. Local industries, including a bleach works and old sludge beds from the adjourning waste water treatment works, put a heavy strain on the land and contaminated the environment. As local industries declined and the area lay neglected, fly-tipping and vandalism further degraded its already damaged environment.

"With the help of Artery and the Mersey Basin Campaign, we saw the potential in these derelict sites," explains Simon Papprill from Stockport Metropolitan Borough Council. "We wanted to unite the sites into one project with one underlying theme: a community park and a nature conservation zone."

Mersey Vale covers a large area on both sides of the river, acting as an important green buffer between Stockport town centre and the large residential area of Heaton Mersey. The area has been described as one of the 'green lungs' of Stockport, and despite its turbid history, nature has been re-establishing its claim on the land, with new woodlands, meadows and wetlands springing to life through the rubble. However, the area was still only being used by the occasional dog walker, horse rider and jogger.

In an extensive community consultation, the Mersey Basin Campaign asked local residents two key questions: what they thought of the area now and what they wanted to happen to the land in the future. The results were unequivocal: people said Mersey Vale was remote, unwelcoming, unsafe and appallingly neglected. Not the sort of place they would choose

to visit. But they were equally emphatic in what they wanted for the future: the dereliction repaired, the site's natural beauty protected and enhanced and its recreational potential opened up.

Over the three year project timescale, Stockport Council have developed close working relationships with local landowners and businesses to secure the land for public use. Key stakeholders from the surrounding area gave invaluable guidance on the project's delivery and ensured that the local community's voice played an integral part of the project process.

Now, children enjoy playing on the grass around Mersey Vale Nature Park's beautiful entrance, which they helped to design through a series of workshops, while parents chat away on the new benches overlooking the river. Access improvements have been made to the Trans Pennine Trail, an internationally renowned coast-to-coast, multi-user route across the spine of northern England, encouraging more walkers, cyclists and horse riders to enjoy the area. Additional footpaths invite people to explore the park further on foot or by bike and many different aspects of nature can be discovered along the way. The new provision for canoeing and fishing allows people to discover and enjoy a river from which they were previously cut off.

Sarah Wallbank, Mersey Basin Campaign's project leader, summarises the achievements: "The creation of Mersey Vale Nature Park has been a fun, interactive and inclusive process. By involving local people throughout the project and by recognising the area's great natural potential, we have created an attractive and sustainable park that the community will look after and enjoy both now and in the future."

NL //
Vuyk Shipyard

A new park on the former Vuyk shipyard brings Capelle
aan den IJssel closer to the river Hollandsche IJssel.
Artery supports locals as they turn towards the water
and facilitates the realisation of their ideas.

The community gathers on the new river banks at the opening of the
Vuykyard // The new access to the river reminds of the former slipways //
The idle shipyard waiting for regeneration

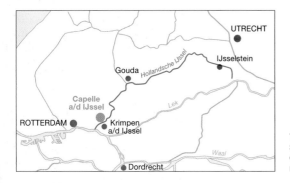

The Netherlands
South Holland
Capelle aan den IJssel
Hollandsche IJssel

The Hollandsche IJssel has always been the artery for the region beyond Rotterdam. Due to its open connection to the port of Rotterdam and thus to the North Sea it is a freshwater tidal river. An open view over the bustling river is a rare but welcome change in this densely populated region.

Within the boundaries of Capelle aan den IJssel it is unique to actually stand on the banks of the river, since most of the riverbank is inaccessible, blocked by housing and other activities. Where once ships were built in a shipyard, today children play in a park. The design of the boardwalk leading towards the water is reminiscent of the slipways on which the ships were once lowered into the river.

"Most goods are transported by road nowadays and the remaining river transport now takes place in ships too big to be built at a small shipyard like Vuyk" explains former owner of the shipyard E.D. Vuyk. When the yard could not be run profitably anymore he decided to close it and sold the land to the municipality of Capelle aan den IJssel. The land remained derelict for many years and though not a pretty sight, the neighbours were happy to use this opportunity to get closer to their river. However, there was no doubt that housing would eventually arise.

When plans to develop the riverbank were presented the neighbourhood was shocked – 123 apartments on such a limited stretch of land. They feared that the unique opportunity to turn this river access point into a lasting recreational space would pass them by. Leen Keij from the local neighbourhood board asked pointedly: "Will Capelle aan den IJssel in future only be referred to as Capelle?"

Concerned members of the neighbourhood board formed the Petit Comité Vuyk and began asking questions around town. By means of questionnaires they collected together numerous ideas put forward by over 300 citizens. After intensive lobbying local politicians granted them a voice in the decision making process.

The different needs of the community, from affordable housing to the advantages of recreational green space, were carefully considered, before the city council altered the land utilisation plan from housing to recreational area.

Financial restraints limited the ambitious project and what should have been an attractive park gradually turned into a modest green. The neighbours saw all their efforts threatened by the possibility that houses would be built to finance the plans for restructuring the area. When the Projectteam Hollandsche IJssel made contact with the Artery initiative, a crucial breakthrough was established. René Kandel, project leader within the municipality of Capelle: "The sheer prospect of support through a European network encouraged the municipality to pursue this project." The committed residents remained doubtful, only when the building board advertising the plans for the park was erected, complete with the European banner and the Artery logo, were they able to celebrate the breakthrough. The point of no return had been crossed – and the plans for the park began to flourish as Artery co-ordinated the continuous participation of the citizens and fascilitated the realisation of their ideas.

The locals owe it to themselves that they can now enjoy the view from a beautiful boulevard along the IJssel River. Grand Café Fuiks offers a safe haven in the neighbourhood all year round. Who knows, maybe one of the riverboats passing below once slid off the ramps at the Vuyk shipyard?

NL //
Baai van Krimpen/
Loswal III

The Hollandsche IJssel was once renowned as Holland's most polluted river. After extensive sanitation it is now ready for rediscovery. Public participation made a real difference in the successful riverside regeneration.

The Baai van Krimpen was not a welcoming place to linger // The river bend at Baai van Krimpen was designated as green space, but was as such hardly used // Former abandoned and neglected site of the 'dukdalf' monument

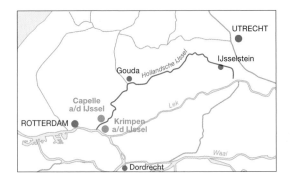

The Netherlands
South Holland
Krimpen / Capelle aan den IJssel
Hollandsche IJssel

A strange construction of massive wooden beams catches the eye on the river bank at Loswal III. It is a 'dukdalf', a mooring post, marking the site where in 1574 during the Dutch Revolt, William of Orange cut the dike and flooded the polder. A bold decision with sweeping effects: it ended the siege of Leiden. More than four centuries later the dikes are long back in place. But after years of neglect the state of the river bend between Krimpen and Capelle aan den IJssel called for action – for the regeneration of the Hollandsche IJssel.

The 'zelling', a stretch of dry land between the dike and the river, had been used for years to deposit sludge. Stuck between the municipal wastewater purification works and a new housing area, the site was appointed as recreational space for its neighbours. The opposite shore, with only a few houses and no industry, offered one of the few open spaces along the river with a precious view into the low lying polder, but little incentive to linger.

Something had to be done. But how to make the right choice? A carefully planned decision by the authorities and everyone follows suit, like in the old days? Once effective strategies were not to be repeated as new approaches were demanded. The successful decontamination of the river bed as part of the 'Project Hollandsche IJssel' offered the opportunity to break new ground in regenerating the riverbanks. Artery facilitated the participation of local residents and invited them to brain-storming sessions. Here they could put forward their ideas and concepts and actively take part in the redevelopment process.

The consultations showed that the free view over the river bend was highly appreciated and had to

be retained. Trees for instance were not wanted. Landscape architects drew on the local's ideas and turned them into illustrative propositions for further discussion. At the Baai van Krimpen the plans of a proposed picnic area with a view on the river attracted the most attention, whereas in Capelle aan den IJssel it was felt that the attractiveness of the monument should be enhanced and a foot path along the river created.

Although a public hearing of residents was not officially necessary, the municipality chose the integration of citizens from a very early stage in the decision making process. Wilco Melenberg of the municipality of Krimpen aan den IJssel explains: "We aimed to obtain the support of the community for our plans as we did not feel like forcing well-meant ideas onto them. In this respect we have had our share of hard learned lessons in earlier projects." Participation improves acceptance and sustainability.

"It is probably the project where we have learned the most", said Albert Koffeman from the Projectteam Hollandsche IJssel. "Preparation is essential," he admits, "be it communities or province, dike reeve or the ministry of water management, we made sure that all authorities involved in the project checked the practicability of the various concepts." Wilco Melenberg adds: "Once you present a concept to the public, you need to be very sure about the feasibility of it."

It was a bold decision to go ahead with the open planning process. Public participation identified an array of possibilities and the involvement of the citizens promises the sustainable regeneration of the Hollandsche IJssel.

NL //
Windlust

The cultural heritage of early prosperity on the Hollandsche IJssel had been long neglected and the windmill at Nieuwerkerk was just a faded memory. Today however, the Windlust turns her sails into the wind once again – not to power nostalgia but to actually work.

13th of May 2006 National Windmill Day: The reconstructed Windlust with its adjoining buildings are again a characteristic landmark // The freshly grained flour is directly processed in the sheltered workshop; the new user of the facilities // The ruins of the Windlust before restoration

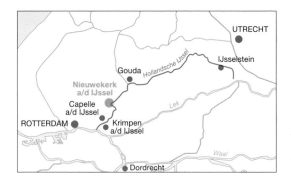

The Netherlands
South Holland
Nieuwerkerk aan den IJssel
Hollandsche IJssel

Convenient water transportation has always attracted trade and commerce to settle along the river banks of the Hollandsche IJssel. For generations windmills have borne witness to the industrious spirit of the region. But with the advent of steam engines and electricity the former power houses and technological predecessors of a long industrial tradition were neglected and many disappeared over the course of time. The Windlust (1741) was almost invisible, only its 'body' survived which was used as a fodder silo for many years. Entrepreneurship saw the potential of this unique relict – as a lift shaft in an office block.

Thanks to the efforts of concerned locals the office block never materialised. Since the remains of the Windlust were still listed as a monument, the former owner feared difficult preservation orders and decided to sell it to the initiators of the 'Molen Kortenoord' foundation. They invested their very own money and the first stone of an ambitious enterprise was laid with the purchase of the derelict 'body'.

It was clear from the start: the historic windmill would be rebuilt. The foundation started to collect money wherever it could be found: from collecting returned deposit bottles at local supermarkets and company component sponsorship, to the symbolic sale of 'IJsselstones' the region's traditional brick stones, which were recovered from demolition sites. A lot was achieved, but not enough – the costly reconstruction and a long-term financing and maintenance plan remained unsolved.

With Artery the venture gained momentum. It initiated a public-private partnership to secure the future of the windmill. "The breakthrough was achieved with Artery", remembers Maarten Molenaar, founder of the action group. "We could now offer a professional concept for the Windlust with long-term utilisation prospects, and a maintenance and financing plan." This inspired confidence in the feasibility of the project and negotiations with building society 'Ons Huis' were finalised successfully – they are now 90 percent owner of the Windlust.

Today the windmill stands in all its glory, sails turning in the wind, grinding corn with its minutely reconstructed machinery. New tenants have moved in, who now make sure that the mill on the riverbank remains appropriately industrious. The Dutch welfare organisation ASVZ Group offers mentally disabled and people with learning difficulties various activities within the unique premises of the resurrected windmill. The cosy brick vaults of the basement house a workplace and bakery. A tea room in the mill annex offers stunning views on the river and a good opportunity to sell the home-made bread.

The ambitious idea of a devoted few, combined with the efforts of reliable supporters, has made this impressive achievement possible. The windmill on the Hollandsche IJssel draws visitors to the river bank again. A new jetty makes it a port of call for the historical sailing ship 'The Helena', and opens up new waters for age old river sailboats.

D //
Barrier-free tourist attractions in the Ruhr Valley

For families with babies, disabled people and seniors a day trip often finishes after just a few metres. Stairs or narrow passageways become insurmountable obstacles. That's not the case in the Ruhr Valley. Here the inter-communal initiative 'Das Ruhrtal' has put barrier free tourism on the map.

Specialised bikes offer disabled people new recreation opportunities // The RuhrtalFerry closes a gap in the Ruhr cycle track // The new information centre at the Harkortsee

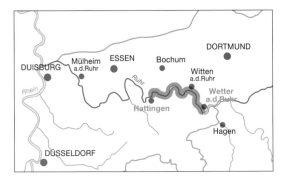

Germany
Ruhr Area
Hattingen - Wetter an der Ruhr
River Ruhr

The Ruhr Valley was once the heart of Western Europe's largest industrial area. But today you'll hardly find any mining or large-scale industry in its place are idyllic artificial lakes and wooded floodplains. The initiative 'Das Ruhrtal' aims to turn the region into a centre for local recreation. To ensure that the region can compete against existing tourist attractions, unique offers for special target groups have to be created.

In line with Artery the Witten Labour and Employment Company (WABE) have developed a unique local recreation attraction in co-operation with regional initiatives. "Especially disabled people find it particularly hard to find suitable jobs and suitable leisure activities. We therefore wanted to interconnect the aspects of local recreation and occupation for disabled people", explains Mona Cartano, a WABE Project Manager.

The pilot project covers several sub-categories. The RuhrtalFerry connects the Hardenstein castle ruins with the Witten-Herbede lock. The ferry operates with pollution-free solar power and closes a gap in the Ruhr cycle track. RuhrtalVelo, the bicycle station in Witten, hires specially designed bikes for disabled people. RuhrtalVelo also gives information on suitable routes and special attractions along the way. Transport for the bicycles is also looked after. RuhrtalMobil transports bikes, canoes, sport equipment and - if desired - also a warm meal for the start or end of the tour or to a specific WABE-managed rest-point.

At Baukey on Harkortsee, one of the three Ruhr reservoirs, an information centre has been set up for local recreation attractions in the Ruhr Valley, in co-operation with the Harkortsee Yacht Club and the city of Hagen. A beautiful half-timbered listed building was especially restored and a disabled accessible boathouse built. Tourists will find a branch of RuhrtalVelo here and also RuhrtalMobil. Water sports are also not in short supply. The Harkortsee Yacht Club is the largest wind sailing club in the region. "Above all we promote sailing for children and young people as well as disabled people with both physical and mental disabilities. Through Artery we now have a suitable infrastructure", says Hans-Joachim König, chairman of the Yacht Club.

The RuhrtalService Organisation operates the individual projects. Half of their employees come from disadvantaged sectors of the job market. Thus Artery has created 20 new jobs for the long-term unemployed, disabled people as well as uneducated young people. "Through this project we expect to see a clear increase in Ruhr Valley tourism. This will create economic growth and thus will lead to more jobs", explains a very pleased Thomas Strauch, Managing Director of WABE.

D //
The River Bath Ruhr in Wetter

Jumping from a rock into cool water. Sunning yourself on a beach in the sun. At the open-air shoreline Ruhrtal natural pool in Wetter the old tradition of swimming in rivers is being revived.

The new River Bath Ruhr before opening. The spacious swimming pool offers many opportunities for leisure and fun right on the banks of the Ruhr // The water of the nature bath is cleaned with a purely biological and physical method – this increases the bathing quality as well as the economic effiency of the swimming pool // The old Ruhr bath facilities needed a fundamental refurbishment

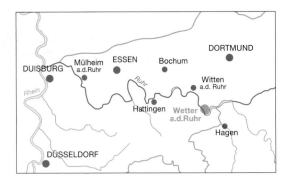

Germany
Ruhr Area
Wetter an der Ruhr
River Ruhr

Just a few years ago it looked completely different around here. Built in 1963 the open air swimming pool beside the lake in Wetter (Ruhr) had deteriorated badly over the years and was in dire need of restoration. The concrete children's pool could not be seen from the sunbathing area. The view on the neighbouring lake, although only metres away, was completely obscured. The pool's chlorine plant was hopelessly outdated. Play equipment, the five-meter jump tower and the double chute were looking noticeably dilapidated. However, the City of Wetter, who would normally have carried the bill for the necessary modernisation and renovations, did not have the money to cover these costs. At the start of the 2004 open-air season, the city cancelled all service contracts, which would have led to the closing of the pool.

After intensive political discussions over the future of the pool and a petition signed by 4,500 citizens from in and around Wetter, calling for the pool to be kept open, the city started looking for a new carrier for the River Bath Ruhr. At the end of 2003 eighteen Wetter citizens set up the 'Our Open-Air Swimming Pool at the Lake Association' (Unser Freibad am See e.V.). At the beginning of 2004 the city finally handed over the pool to this new operating society, for the next 25 years.

Now the question of the pool's urgently needed modernisation arose. The operating society developed plans for the conversion of the renovation in a public-private partnership with the Regional Association Ruhr (RVR) and the City of Wetter in the context of the Artery project. From the chlorine-pool, a natural pool on the Ruhr would emerge.

"With the River Bath Ruhr in Wetter we wanted to reinvent traditional public river swimming pools along the Ruhr", explains Torsten Göse, president of the operating society. Public river swimming pools were far more common along the Ruhr until the beginning of the 60's. One of these was the Ruhr Bath in Wetter. However, the increase in water pollution led to the open river baths being converted into closed chlorinated open-air swimming pools. This is what happened in Wetter. "With a natural pool with a direct view on the Ruhr we want to reintroduce locals to the river and make it part of their lives once again", says Frank Bothmann, Project Manager with the Regional Association Ruhr.

The cleaning of the water in a natural pool takes place in a purely biological and physical manner, without chemical disinfectants such as chlorine. Which not only benefits the natural pool itself, it also lowers operating costs substantially. In Wetter a so-called 'Neptunfilter' with a surface of 1,300 m² ensures the cleanliness of the water. The natural pool has capacity for 2,200 visitors.

But it's not only the pure water that gives visitors the feeling of bathing in a natural environment. A new terrace on the river promenade has reopened the view of the Harkotsee. A waterfall feeds the large pool with clean water. In the non-swimming area of the large pool a water current leads into a natural looking grotto. In place of the jump tower there is now a jump rock. From a beach chair on the sandy beach parents can now watch their children in the plunge pool or in the water playground.

D //
River Adventure
Station

Can fish fart? Do ducks get cold feet in the winter? Do fish sleep in beds? And how do they manage to stay under water? Visitors can find the detailed answers to questions like these by visiting the Museum Ship, anchored at Mannheim.

Children learning about the aquatic habitat of the Neckar and river navigation // The historical paddle steamer now houses an interactive river exhibition

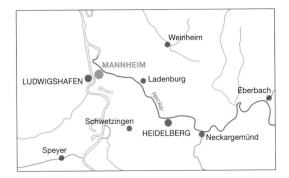

Germany
Metropol Rhine-Neckar Region
Mannheim
River Neckar

Since 1986 a historical paddle steamer has been anchored in the middle of Mannheim's city centre. The ship has belonged to the Baden-Wurttemberg State Museum for Labour and Technology (LTA) since 1990. The permanent exhibition takes visitors through areas such as the engine room and bridge and shows original ship's models that tell the story of the history of inland waterway craft. There are also special exhibitions which focus particularly on nature topics.

In the context of the Artery project, the idea to organise an exhibition based upon the Neckar on board the museum ship, was developed by the neighbouring federation Heidelberg-Mannheim. 'Was(s)erleben'(Water experience/water life) opens discussions to encourage ecological awareness. The Neckar should once again be a place for nature, leisure and recreation. In the last few decades the local community has moved away from the river, "Our goal is to reintroduce the local residents to the Neckar and its ecological system", says Ruben Scheller, Project Manager of the Nachbarschaftsverband (Neighbourhood Association) Heidelberg-Mannheim.

The museum ship is designed as an out of school learning centre. The exhibition is therefore addressed particularly at children between six and fourteen years. On 'Discovery Islands' children can investigate the flora and fauna in and around the Neckar through a magnifying glass. In the next area they can experience how the river has changed over time: from a wild flow to a canalised waterway. Touching and observing are the crucial criteria here. In feeling boxes different types of river rock can be discovered. Flow models

with wind and sand or with wind and water illustrate complex topics such as erosion and accumulation, sliding and undercut slopes. A periscope introduces visitors to a further Artery project: visitors can look at a re-naturalised piece of Neckar bank on the other side of the river (see following page). River landscapes form the central theme of this part of the ship.

Further towards the back of the ship visitors can learn about life in the water. A Cartesian diver explains to children, exactly why fish can sink and rise in the water. DVD animations show reports on freshwater fish and water birds. At the 'no stupid question wall' questions are posted such as: "Can fish fart?" or: "Do ducks get cold feet in winter?" The answer is given by a plastic fish, which the children catch out of a tub.

On deck visitors can find out all about navigation and technology. Here they can tie knots and even build bridges – and test their stability straight away. The highlight of the exhibition is a four-metre long model of a lock. Pupils can navigate self-made ships through the sluice from one basin to the other. A tutor is always available to answer questions regarding the exhibition and individual exhibits.

Pupils can not only approach the river in a playful capacity, but also scientifically. In the Neckar laboratory twelve microscopes and a zoom lense microscope with inserted video camera are available. "With river education we try to bring the Neckar to the children and with it the future generation closer to their habitat", explains Josephina Kaiser-Heinstein from the Baden-Wurttemberg State Museum for Labour and Technology.

D //
Agendapark Living Neckar

Artery boosts Neighbourhood Association's efforts
to create a green stretch along the river Neckar.
Between Mannheim and Heidelberg an inter-communal
co-operation has resulted in a significant transformation
to the river's landscape. Improvements benefit both
people and wildlife.

On the new playground children can experience the element water // Artistic
landing stage at the new town entrance of Ladenburg attracts visitors to the
river bank // This monotonous riverside in Mannheim is turned into a new
shallow water zone

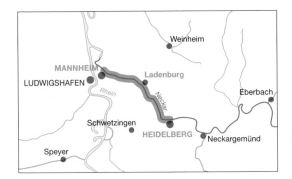

Germany
Metropol Rhine-Neckar Region
Mannheim-Heidelberg
River Neckar

Local recreational areas have gained greater importance over the last few years. They create a balance for people who spend most of their day indoors. This can pose particular challenges in densely populated areas.

With Artery, the Nachbarschaftsverband (Neighbourhood Association) Heidelberg-Mannheim, or NV, an association of 18 cities and municipalities in the Rhine-Neckar region, and the Verband Region Rhein-Neckar (Regional Spatial Planning Association Rhine-Neckar) joined forces to face this challenge. Their project 'Agendapark Living Neckar', aimed to renaturalise the river and create new recreational resources.

"The thriving Rhine-Neckar region is home to global companies, industry, culture and science. It borders the beautiful Palatinate and the Odenwald forest. But locally there were no green open spaces for the residents to enjoy, even though they had the river right on their doorstep", explains regional expert Klaus Mandel.

Along the 20 kilometer stretch from Mannheim to Heidelberg, tall bank reinforcements and overgrown embankments often left the river inaccessible and invisible. It was therefore a major goal to improve river access.

In Mannheim's city centre the Neighbourhood Association has initiated a 100 metre long shallow water zone. People shall reach the water easiliy, sit on the shore and observe the aquatic habitat. It will be possible to walk knee-deep into the water where before you could not even touch the surface.

Ladenburg had a similar problem. Wild overgrowth completely blocked the embankments. Now a new town entrance faces the Neckar. Gently sloping lawns and paths run down to the waterline inviting children to play and adults to linger. More attractive riversides were also established in municipalities like Edingen-Neckarhausen and Dossenheim.

Promoting the green stretch as a large recreational area, people were made aware of the cycling paths that run along the Neckar connecting the communities between Heidelberg and Mannheim. New signposts now point out these walking and cycling routes.

While industrial infrastructures cut people off physically; urban developments also led to people losing touch with the river. It was therefore vital to strengthen the population's awareness of the Neckar.

In cooperation with educational institutes like the WBW, educational concepts were designed. A 'river teacher' offers exciting events focussed on culture and nature along the Neckar. Training programmes for teachers promise long-lasting effects. A water playground in Heidelberg adds to the children's familiarity with the liquid element. These initiatives will not only change leisure time habits, but also raise environmental awareness.

The canalising of the Neckar forced local wildlife into retreat. Natural habitats were destroyed or blocked. Artery helped to start to reverse these consequences. Re-naturalising projects, like the creation of shallow water zones or alluvial forests, give local wildlife the chance to flourish and, like the European beaver (Castor fiber), resettle in their indigenous habitats.

"Many individual partners contributed to the 'Agendapark Living Neckar' project. The combination of enhancing the riverside and raising the public's awareness is very special", states Ruben Scheller, project manager of the Neighbourhood Association. This combination ensures the sustainability of the Neckar project.

D //
Wernau Neckar-River Reserve

The re-development of part of the former riverbed makes a valuable contribution to the conservation area on the South bank of the Neckar in Wernau. Further projects aim to re-naturalise the heavily environmentally degraded river and to create new recreation areas.

This small brook will be turned into a fish pass // With opening the Neckar dike the Erblehensee gets reconnected to the river // The connection with the Neckar turns the former lake into a stream

Germany
Greater Stuttgart Region
Altbach and Wernau
River Neckar

Ecologically the river Neckar in the Greater Stuttgart Region has become extremely strained. In the heyday of urbanisation and industry it served as a convenient 'tool': used and altered as needed.

The greatest interruption to the river's natural flow took place in the 1960's. As a means of flood protection the Neckar was straightened and a dike built. Old tributaries were cut off and a new riverbed reinforced. The repercussions of these developments affected the natural riverside habitat in many ways.

Now, forty years later, the Artery programme supports planned regional initiatives to re-naturalise parts of the local embankments. The municipalities of Wernau, Plochingen, Altbach and Deizisau have worked on an all encompassing concept – the 'Green Initiative Neckarknie' ('Grüninitiative Neckarknie') – to counteract some of the alterations made.

Project 'Erblehensee' (Lake Erblehen) reconnected the river with part of the original riverbed. The lake was a remnant of the canalisation of the Neckar close to the town of Wernau and an important conservation zone. In the context of Artery the Neckar dike was lowered and in two places opened. Now, fresh river water constantly flows into the former stagnant pool. "We want to return parts of the Neckar to its natural state and undo some of the destruction of the past", explains Dieter Bayer, from the regional council Stuttgart (Regierungspräsidium Stuttgart) and project manager of the work group 'Erblehensee'.

This has had immense ecological benefits. The newly created sidearm for example, offers a retreat for aquatic fauna from the fast flowing stream especially during high water, and provides a nursery for fish.

The reconnection of the Erblehensee is an important revaluation of the already sensitive conservation zone and one step towards a natural riverside along the Neckarknie. "With the participation of Artery it was possible to put riverside regeneration in the foreground of regional strategies" praises Konrad Störk from the water management department of the regional council. "Artery was an impetus to finally get started." Artery is also involved in two follow-up projects.

In Altbach a feasibility study for a fish pass has been successfully established. A large dam in Deizisau hinders aquatic animals from moving upstream. As a consequence many fish cannot reach their spawning grounds and the diversity of the fish population is threatened.

The solution is a fish pass. Just before Deizisau an old tributary of the Neckar bends off around the EnBW power station into the Heinrich-Maier-Park. Nearby a water cooling channel for the power station leads into the river on the other side of the dam. A connecting brook between the old tributary and the cooling channel would enable all water animals to bypass the insurmountable obstacle in the river.

The realisation of the fish pass not only increases the ecological value of the riverside, but also the attractiveness of the Heinrich-Maier-Park. With Artery plans were made to connect this park with the green spaces in Deizisau. These plans envision an aesthetically appealing footbridge rising in gentle slopes over the Neckar. The special construction, which avoids high gradients, allows a safe crossing for bikes and wheelchairs.

Text: Rudolf Kerndlmaier

Transnational Exchange and Transfer of Knowledge

The Artery partners celebrating the first Project Group Meeting in Gouda

Introduction

The North West of Europe is a dynamic and prosperous area with common social and economic pressures. Where globalisation and European integration reveal the interdependencies between regions, economic dynamics demand competitiveness. Where most benefits are gained from mutual interaction on various levels, the importance of increased cohesion becomes apparent. The central aspect of cohesive development is co-operation. The INTERREG IIIB programme is build with this idea of co-operation in mind. It increases connections within a European network. Within this network both personal and institutional relations facilitate the exchange and creation of knowledge.

The co-operation of cities and regions on a European scale, therefore, benefits the development of all partners involved. A higher degree of co-operation leads to sustained and balanced development and territorial integration. In a cross-sector approach, experiences are exchanged on all levels. Europe becomes tangible in this cross-border co-operation, strengthening ties and multiplying benefits.

Co-operation in Artery

Artery is a project within the INTERREG IIIB programme and combines the efforts of 16 partners in 5 regions. Its aim is to set benchmarks for riverside regeneration by creating sustainable environments in areas that have suffered from the effects of post-industrialisation. INTERREG has revealed the potential for interregional co-operation. It recognizes both the characteristic differences of the regions and the similarities of their challenges.

All Artery partners had already defined long-term development concepts to tackle local challenges. The different approaches used in terms of spatial planning, process organisation, target group involvement and local co-operation, led to expertise in different fields of knowledge. The partners collected different experiences, making each one strong in specific aspects. Following the approach of expert regions and learning regions, Artery offers a platform where methodological strengths can be exchanged. The partners co-operated in Artery as one European project, learning from different experiences in daily collaboration. Artery used the methodological approaches of the INTERREG programme to organise its interregional co-operation and aims to improve the results of knowledge transfer. Common strategies to improve projects were developed from the combined knowledge of the partners.

Artery is based around four common themes: public participation, regional strategies, public awareness and public-private partnerships. These themes are designed to facilitate exchange, increase co-operation between the regions, and mutually enhance the project partners' existing knowledge. Transfer of knowledge and examples of best practice effectively transformed the pilot projects and resulted in a more cohesive, balanced, and therefore more sustainable development, maximising the project benefits.

Programme Secretariat Julia Eripret, INTERREG IIIB NWE Secretariat ENO, Lille

Leadpartner Frank Bothmann, Regionalverband Ruhr, Essen

Project Coordinator
Holger Platz, PLANCO Consulting GmbH, Essen

Financial Manager
Gunnar Platz, PLANCO Consulting GmbH, Schwerin

Ruhr Region	Mersey Basin	Hollandsche IJssel
Regional Leadpartner Frank Bothmann, Regionalverband Ruhr, Essen	**Regional Leadpartner** Sarah Wallbank/Iain Taylor, Mersey Basin Campaign, Manchester	**Regional Leadpartner** Albert Koffeman, Provincie Zuid Holland, Den Haag
Thematic Leadpartner Claudia Wolters, Regionalverband Ruhr, Essen 'Regional Development Strategies'	**Thematic Leadpartner** Sarah Wallbank, Mersey Basin Campaign, Manchester 'Public Private Partnership'	**Thematic Leadpartner** Albert Koffeman, Provincie Zuid Holland, Den Haag 'Public Participation'

Ruhr Region

Pilot Project Partners
Pilot: River Bath Ruhr
Bernd Hagedorn/Torsten Göse, Trägerverein 'Unser Freibad am See Wetter (Ruhr) e.V.'
Manfred Sell, Stadt Wetter

Pilot: Accessible Tourist Facilities in the Ruhr Valley
Mona Cartano/Thomas Strauch, Wittener Gesellschaft für Arbeit und Beschäftigungsförderung mbH, Witten
Klaus Tödtmann/Gisela Tervooren, Ennepe-Ruhr-Kreis, Schwelm

Mersey Basin

Pilot Project Partners
Pilot: Stockport Riverside Park
Sarah Wallbank, Mersey Basin Campaign, Manchester
Simon Papprill/Vanessa Brook, Stockport Metropolitan Borough Council, Stockport

Pilot: Speke Garston Coastal Reserve
Iain Taylor, Mersey Basin Campaign, Manchester
Tom Workman, Liverpool Sailing Club, Liverpool

Hollandsche IJssel

Pilot Project Partners
Pilot: Windmill 'Windlust'
Albert Koffeman, Provincie Zuid Holland, Den Haag
Maarten Molenaar, Stichting Molen Kortenoord, Nieuwerkerk a/d IJssel
Ferry Warnaar, Gemeente Nieuwerkerk a/d IJssel

Pilot: Vuyk Shipyard
Albert Koffeman, Provincie Zuid Holland, Den Haag
René Kandel/Leen Verschoor, Gemeente Capelle a/d IJssel

Pilot: 'Baai van Krimpen Loswal III'
Albert Koffeman, Provincie Zuid Holland, Den Haag
René Kandel/Leen Verschoor, Gemeente Capelle a/d IJssel
Wilco Melenberg, Gemeente Krimpen a/d IJssel

Key Benefits of
Transnational Exchange in a Community of Practice

~ Tackles issues beyond national borders

~ Generates transferable models for different regions

~ Keeping to one schedule speeds up process

~ Sharing of expertise and development costs

~ Increases connections in European networks

~ Capacity building for organisation and staff

~ European dimension to local project

~ Co-operation leads to competitive regions

Artery organisational structure

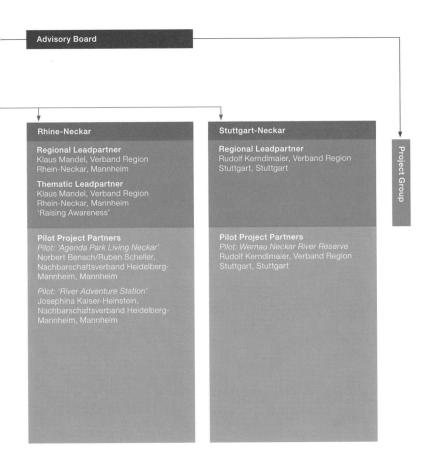

Advisory Board

Rhine-Neckar

Regional Leadpartner
Klaus Mandel, Verband Region
Rhein-Neckar, Mannheim

Thematic Leadpartner
Klaus Mandel, Verband Region
Rhein-Neckar, Mannheim
'Raising Awareness'

Pilot Project Partners
Pilot: 'Agenda Park Living Neckar'
Norbert Bensch/Ruben Scheller,
Nachbarschaftsverband Heidelberg-
Mannheim, Mannheim

Pilot: 'River Adventure Station'
Josephina Kaiser-Heinstein,
Nachbarschaftsverband Heidelberg-
Mannheim, Mannheim

Stuttgart-Neckar

Regional Leadpartner
Rudolf Kerndlmaier, Verband Region
Stuttgart, Stuttgart

Pilot Project Partners
Pilot: Wernau Neckar River Reserve
Rudolf Kerndlmaier, Verband Region
Stuttgart, Stuttgart

Project Group

Transfer of Knowledge

Knowledge can be defined as the combination of information, competence and skills used to solve a problem. It is more than plain information and sheer accumulation of data. Knowledge can be generated only by an individual assessment of data. It is a process not a status, and new knowledge can be generated through exchange with already existing knowledge, creating networks that eventually lead to open discussions and collective decisions.

Knowledge and experiences are considered an important resource to any project. However, this is a source usually confined to a single pilot. In Artery this knowledge and experiences were made available to other projects in the exchange. The four common themes of the Artery project are a means to facilitate the transfer of methods trans-regionally and trans-nationally – turning Artery into a learning partnership.

Every organisation holds a wealth of knowledge. But in order to tap into this pool it is necessary to identify the form in which it is contained. An organisation holds both implicit and explicit knowl-edge. Implicit or tacit knowledge is confined to a person and consists of personal experience and skills. It is internalised and as such a person might not be aware of their own knowledge. It is hard to articulate and therefore difficult to share with others. Explicit or documented knowledge on the other hand, is expressed, verbalised and easily transferable in the form of data. For an organisation to benefit from the valuable experience of its members, it is necessary to be able to transform implicit or tacit knowledge into explicit knowledge, to make it available to others.

Within Artery the personal contact between practitioners and other knowledge holders was the motivation to articulate experiences. In exchange with others, individual knowledge is revealed, when contrasted with different opinions and experiences.

An important aspect of successful knowledge transfer is communication. It is facilitated by precisely defining key concepts, especially when co-operating with people from different professional backgrounds. Knowledge starts with calling things by their name, it

Knowledge Resource Networks

Knowledge networks function on three basic principles:

~ transfer of information between partners

~ equal access to communication and resources

~ collective activities to establish reliable personal bonds between partners

Knowledge Transfer

Knowledge starts with having the same common understanding of things. It is, therefore, important to:

~ Identify common themes of a partnership as transferrable knowledge

~ Identify the key holders of this knowledge within the partner organisations

~ Motivate them to share their knowledge within the transnational network

~ Design mechanisms to transfer this knowledge effectively, like communication routines, workshops, regular meetings or accessible, internet-based data

~ Execute the transfer plan by keeping a binding schedule

~ Measure results – regular update reports on the implementation of individual common themes ensure successful transfer

~ Apply the knowledge transferred

is therefore recommended to compile a glossary at the start of any programme.

Expert Regions in a Learning Partnership

According to their previous experiences, four of the regional partners were appointed thematic leadpartners for a specific common theme. They were responsible for the promotion and distribution of their expertise at both a regional and interregional European level. The regional partners are the essential connection between the local and international partners; they represent Artery in the region, while representing their regions and the pilot partners within Artery. This subsidiarity idea ensures that decision making stays as close to the pilots as possible, while the interconnectedness of the regional leadpartners makes them the knowledge focal point. They find themselves at the hub of information transfer between pilot partners in the region, as well as in the European context.

The organisational structure of the Artery project was set up in order to facilitate the transfer of knowledge. A range of different tools was used by the Artery partners, such as workshops, regular meetings and a system of different reports, as they proved to be helpful in keeping all partners informed. They create opportunities for communication and the exchange of ideas.

Thematic Leadpartners

In their function as thematic leadpartner, regional leadpartners prepare the pilot input for transfer. They guide the compilation of available experience as explicit knowledge and provide technical and financial support. They also provide advice to the pilot partners based on knowledge gained in project group meetings. A comprehensible methodological approach bring together relevant experiences in the respective common themes, providing a theoretical background.

While the theoretical input from seminars and papers was much appreciated by the partners, they found the

Difficult Transfer

The Artery partners expected to profit from each other's regional networks of public-private partnerships. International connections with companies and enterprises were therefore evaluated accordingly. The reality however, proved to be rather different, as potential partnerships depend very much on individual and personal relationships. It became clear that only the 'methods' for establishing regional networks of public-private partnerships could be transferred. Personal contacts and regional networks could not help other regions in eliminating the acquisition process.

live example of the pilot projects more beneficial. The huge experience from MBC in the field of public-private partnerships for example, inspired the Dutch and German partners, who had little or no previous experience in this field, to develop new project management processes.

Project Group Meetings

The regional leadpartners meet regularly, as together they form the project group, the main co-ordinating body of Artery. They report to the project group and chair discussions of their common theme. Minutes are used to collect reports from the partners concerning the implementation status, these allow all partners to learn from other experiences and follow up on further developments. Knowledge about best practice and avoidable stumbling blocks encountered in the pilots, can be transferred.

Transnational bonds: the partners pouring water from the rivers Hollandsche IJssel, Ruhr and Neckar into the Mersey

The successful realisation of the ambitious Windlust pilot often attracted the partners' attention during project group meetings. The reconstruction of the neglected and almost non-existent historic windmill and its adjacent buildings into a day-care centre for disabled people, inspired the initiators on the Ruhr to look for new possibilities within their projects.

Exchange between Pilot Partners

The pilot projects are the actual core of Artery. Hands-on experience is collected during the implementation of the pilot's common themes. Strategic objectives meet reality and sometimes need to be adapted to local circumstances. Reported back through the leadpartners, it is valuable feedback and helps to create more appropriate solutions for the other pilots.

Personal visits between pilot partners can help to increase the mutual understanding of achievements, and sometimes point out new opportunities. The initiators of the River Bath Ruhr paid visits to the Hollandsche IJssel and together with the partners of the Windlust they looked at new ways of public-private partnerships and better use of public relations. From this exchange both the windmill, as well as the river bath benefited from new ideas and fresh input.

Workshops

Throughout the course of Artery, workshops have been held on different aspects of the common themes. Organised by the respective leadpartner, they have enabled participants to bring their knowledge to comparable standards, making room for discussions to work out a consistent understanding of the concepts.

Following the public participation workshop, the Stuttgart region benefited from the knowledge of the Dutch partners and were inspired to look into the possibilities of new partnerships for its Neckar redevelopment scheme. Their useful experiences of public participation lead to the establishment of an interdisciplinary workgroup with local authorities, businesses and the general public at the Wernau Neckar River Reserve. This process of public participation identified the need for better access to the redeveloped site which was then integrated in the further design of the reserve.

Common Themes Reports

These reports were compiled by the thematic leadpartners. They summed up the valuable knowledge accrued by the partners, and collected the hands-on experiences from the different pilots. This established a guideline and principal direction for the respective common theme work.

The well-documented experience of the Mersey Basin Campaign and its efforts to involve businesses in the regeneration of riversides, encouraged the Dutch partner to look for similar opportunities in their own projects. Such partnerships increase the social and environmental responsibility of local businesses and help promote sustainable projects. These were the driving aims when approaching investors for a new café restaurant to be built on the former Vuyk shipyard. This resulted in a successful partnership with Fuyk's in maintaining the newly developed site.

Status Reports

Standardised forms were used to collect comparable information from the different pilots. Sent out by the thematic leadpartner to each learning partner, they asked questions about the status of implementation of a common theme. This allowed for progress evaluation and indicated where additional support might be needed.

Sustainable maintenance is a key aim for the restored windmill Windlust on the Hollandsche IJssel. The innovative solution found by the 'Molen Kortenoord' foundation is being used as a blue-print for the creation of a new organisational structure for the Speke Garston pilot. A management company

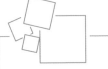

Obstacles and Challenges in Knowledge Transfer

Some factors to keep in mind:

~ Distance between partners – a lack of shared social identity

~ Language barriers and cultural differences

~ Different areas of expertise and therefore a lack of common knowledge base

~ Internal conflicts – for example, professional territoriality

~ 'Hidden' or intuitive competences (tacit knowledge) might go unnoticed

~ Visual representations to transfer knowledge are not always available

~ Misconceptions through faulty or misunderstood information

~ Motivational issues

was set up for the project in Liverpool, which now separates its responsibility as supervisor from operational obligations.

Transferability

The evaluation of the transferability of knowledge is of great concern when working together in an interregional project. In order for it to be used for other projects, transferable knowledge needs to be deductible from specific experiences.

Artery partners at the Public Awareness workshop in Wetter

Success Story: Ruhrtafel Circle Every year the Mersey Basin Campaign (MBC) and the Northwest Regional Development Agency host the renowned Northwest Business Environment Awards. At a prize-giving ceremony held in Manchester, the Northwest's most environmentally friendly companies are honoured.

During the Artery programme the MBC invited the European partners from the Hollandsche IJssel, Ruhr and Neckar to this gala event. Fascinated by the ceremony's stylish atmosphere, the Ruhr partners saw the potential of such an event for establishing public-private partnerships in the Ruhr Valley. It would be the ideal way to start a regional business network. Therefore, after the event the partners sought a way to transfer the impressions gained from the Northwest Business Environment Award to the Ruhr Valley.

Inspired by the MBC's culture of public-private partnerships, the Ruhr Valley regional partners succeeded in bringing together businesses in an informal 'Ruhrtafel' Circle meeting. One outcome of this has been extra financial support for new leisure activities in the valley area, including the Ruhrtal Camp 2005, a children's camp which attracts young people back to the river.

In the common theme of public awareness high hopes rested on plans to tour with the river exhibition from the Neckar to other Artery partners. Though transferable in its general concept, the touring river exhibition never materialised. The contents of the exhibition were too much adjusted to the peculiarities of the Neckar, making it non-transferable. The underlying idea of explicitly addressing young school children however, was a concept eventually taken on by several other partners. The initiators of the Windlust for example, started an educational programme about the tradition and technical features of historical Dutch windmills.

Collective Timeline

The principle of subsidiarity allowed the pilots to implement their projects with greater independence. But the regulations outlined in the INTERREG IIIB programme required that all projects to stick to a common timeline. An action plan was set up to meet these requirements. This timeframe and its resulting pressure speeded up the local decision making

Organisation

~ Chart a transparent organisation structure and ensure that the partners are familiar with it

~ Clearly define responsibilities in a set of rules

~ Review the status of implementation at different stages of the project

~ Make sure that the partners dedicate an adequate level of staffing

Controlling

~ Maintain separate financial records for each project

~ Call in professional financial administrators to assist project managers

~ Install a financial controlling system

~ Bear the completion deadlines in mind when scheduling activities

~ Keep track of whether expenditure is proportional to the progress the projects are making

~ Keep a sharp eye on the timeframe for implementation and payments

processes. The idea of joint implementation is not a 'one-size-fits-all' approach, but rather a step towards a community of practice.

An important corner stone is a contractual agreement, the Joint Convention. It determines the partners' individual responsibilities and forms the basis of successful co-operation. It is a prerequisite of the programme's joint secretary in the process of reimbursing expenses. Together with the timeline it helps to prevent and minimise delays. The status of implementation for each pilot had to be reported to the same joint secretary. This is a procedure not to be underestimated – as an external authority it increases the reliability of the partners and benefits the communal work.

Funding

Artery allowed small and medium-sized projects to apply for funds from the INTERREG IIIB initiative, these funds would not have been available for the single pilots. The organisational structure of the Rhine-Neckar pilots can serve as a model for the integration of various sub-regional partners.

Funding through Artery often initiated a project's first steps, and while the pilots were getting started they attracted additional funds – often serving as cata-

lyst for various projects in the vicinity of a pilot. The flexibility in allocating funds within Artery allowed for an important improvement of the total programme.

Communication

In order to generate an identity as a European project, communication is a determining factor of its success. Artery is a project of ten pilots in five different regions – it takes effort to create a feeling of solidarity and unity. Shared communication with a recognisable logo on all media, newsletters and news from partner regions in local media, creates a unified identity.

At the beginning of a project, communication is needed to raise public awareness. The fact that regional pilots were part of a greater European project was a unique feature in many regions and made interesting news. Extensive coverage supported the creation of a common basis for riverside regeneration. Projects that had reached completion were spotlighted again by journalists eager to report on their results. Communication clearly helps to involve people in a project. Ultimately, as the examples in the common

'Spin-Off' through Knowledge

Transfer of knowledge within the transnational co-operation of the Artery initiative helped partners qualify to join other European programmes. Participating in the INTERREG project helped the Witten Labour and Employment Company (WABE) to gain access to other programmes and ensured another 80,000 euros was approved for its project.

Basic Rules of Co-operation

~ Define common themes as key concepts early on and agree on a common glossary

~ Networks rely on the partner's investment of time and energy in order to be built and maintained

~ Be open to new experiences and learn from partners

~ Stay clear of inappropriate adoption of generalised methods, work towards a learning partnership with situative knowledge transfer

theme work of awareness and public participation have shown, this determines how effective and sustainable the Artery projects are.

A communication plan was made for the Artery project. The Artery website contains information about each individual partner and its respective projects, and highlights the common themes, tying these pilots together in their shared approach. A regular newsletter gives in-depth information about a single pilot, while giving an update on its general progress. The European scope of their projects was also part of the individual partners' publications. 'Het Peil' of the Projectteam Hollandsche IJssel, as well as 'Source' of the Mersey Basin Campaign often contained references to the Artery partners. Press releases about the pilots carefully integrated the transnational aspect, attracting more interest among local media. Newspaper articles frequently report on the European scope of Artery pilots enhancing their image in the region.

Benefits

A successful transnational partnership is greater then the sum of its parts. Artery has effectively changed the outcome of individual projects, as it has shed light on previously undetected similarities in partner pilots. Even though in the beginning transnational co-operation appeared to be only a minor factor at project level, it was nevertheless one of the grant stipulations. Networks were established as the confidence between the partners grew and these will continue to act as a stimulus for future interregional projects.

The Artery effect has been felt beyond the pilot projects, too. Acting as a catalyst for various other schemes, it has helped to trigger new local initiatives. At the same time it supports new and stronger regional partnerships and networks. In all, over 13 million euros in extra funding has been levered in by the pilot projects. This is a direct result of the catalytic effects that the Artery project has had, and the way in which new and existing networks have worked together to foster new partnerships.

Co-operation in Artery as a strategic network was a prerequisite for the successful implementation of the ten pilots and resulted in an alliance for sustainable riverside regeneration.

All European partners gathered for the Artery launch event at Hattingen in 2004

Text: Albert Koffeman

Public
Participation

The opening of the Vuykyard: culmination of the public participation process in Capelle // Early planning stage in Stockport

Introduction

Rivers and riversides are recognised as an integral part of a region's landscape and socio-economic well-being - after decades of neglect, riversides have become a focal point of public attention. Waterfronts are increasingly regarded as assets, and therefore they are often part of regional development programmes. In order to attract the public back to formerly neglected sites, project managers responsible for these programmes need to listen to the public.

Ambitious development projects have failed in the past, because they lacked acceptance from the local population. An inadequate understanding of a region's identity often leads to inattentiveness for the specifics of local requirements. This is a major reason for the lack of acceptance of newly developed sites. Creating public spaces using the hierarchical top-down approach of traditional decision making processes can often result in unsustainable decisions. Public space is a vital part of people's daily environment, and is therefore subject to public scrutiny. Its planning should therefore be brought closer to the people.

If the public have been asked for their opinion and their needs have been addressed, a higher acceptance level is generated. Neglecting the interests of the general public however, can lead to the gradual decline of a public space into a derelict no-man's land. If plans fail to address local needs or insufficiently solve the targeted problems, their implementation can trigger opposition. This can delay improvements and even lead to vandalism.

The insights gained in the course of the Artery projects have demonstrated the advantages of public participation. This chapter will outline these advantages and illustrate how they can help to achieve high quality projects with improved sustainability.

What is Public Participation?

Public participation is the process of actively including the public in the development of policies and projects affecting their community. On the one hand, it provides the public with information and instruments that enable them to articulate their interest in a development process, on the other, it helps project managers understand the needs of those who have an interest or stake in the outcome of a project, and enables them to develop alternatives that will ensure the project's sustainability.

Public participation is a mutual learning process that requires a will to share control and responsibilities. It allows the general public to become involved in different levels of the planning procedure; from basic information supply to self-determined and active participation in the work process. Carefully managed, it is a valuable project planning tool, as it helps to identify possible concerns within a community. By successfully involving the public, the community can take on responsibility for the project, and a feeling of ownership is generated, which can make the project more sustainable in the long run.

Public Participation

The Role of Public Participation in Artery

Public participation is a useful method, which can assist in a project's development. This is why it was chosen as one of Artery's four common themes. Within the Artery programme it serves as a common strategy to increase public ownership and generate wider support for a project. The different degrees of experience amongst the European Artery partners made the potential benefits of mutual knowledge transfer apparent. Artery chose a leadpartner for every theme to manage this transfer of knowledge, strengthening the transnational co-operative aspect of the Artery initiative.

The Projectteam Hollandsche IJssel was appointed as Artery leadpartner due to the experience they had gained during the ambitious Hollandsche IJssel redevelopment scheme. It was in charge of the promotion and facilitation of public participation as a common theme. Through workshops and reports the Projectteam provided the learning partners with advice for the implementation of participatory techniques, opening up new possibilities for the German and English projects. The project – with the motto 'schoner, mooier, hollandscher' which translates into 'cleaner, prettier and more characteristic' – is an initiative aimed at redeveloping the Hollandsche IJssel with a combined environmental, aesthetic and cultural approach. The area covers over 20 kilometers between Gouda and Rotterdam and comprises 13 local authorities including municipalities, the Ministry of Transport, Public Works and Water Management and the province of South Holland. Over the course of the project, which started in 1996, the Projectteam collected practical experience from different forms of participatory processes. This served as motivation to promote public participation in other

Public Participation from a European Perspective

Over the last few decades public participation has been of growing importance in international legislation:

~ **Rio 1992** The Rio Declaration and Agenda 21 of the United Nations Conference on Environment and Development (UNCED), promote a common strategy for sustainable development, taking the ecological, social and economical factors of development into account. Principle 10 of the Rio Declaration stresses the need for residents' participation in environmental issues.

~ **Aalborg 1994 and Lisbon 1996** The Aalborg Charter of European Cities and Towns towards Sustainability was signed in 1994 by 80 communities who joined forces to face the challenges of Agenda 21. With the Lisbon Strategy two years later the group agreed on a Local Agenda 21.

~ **Aarhus Convention 1998** This agreement between the United Nations Economic Commission for Europe (UN ECE) and the European Union is an elaboration of the Rio Declaration. It provides residents with the right of access to information and participation in environmental decision-making processes. It is an important step towards environmental democracy.

~ **EU-Water Framework Directive (WFD)** The Water Framework Directive of 2000 contains provisions to secure the quality of all forms of water, setting strict standards that have to be met within a specific time-frame. Great emphasis has been placed on the involvement of the public to achieve these objectives, and to make it a binding element for all aspects of the policy.

Artery pilots. The Projectteam encouraged residents to participate in the improvement of the Hollandsche IJssel. In reference groups residents were given the opportunity to voice their concerns and give advice.

Through the Projectteam Hollandsche IJssel the Artery partners were able to draw upon long-term experience with public participation in the Netherlands. The first humble beginnings of public participation in the Netherlands began with a law regarding the publication of governmental documents in 1980. This law requires governmental institutions to provide access to the majority of their documents, opening administrative procedures to the interested enquiries of the public. The foundation of active public participation was laid with the law on environmental impact assessment Milieu Effectrapportage M.E.R. Since its introduction in 1987 it has prescribed research into a project's effects on the environment. This process offers the opportunity for residents to voice their concerns about a planned project. These concerns have to be heard and included in the final report for further assessment by the authorities. It is an obligatory pro-

cedure for any plan or project with environmental consequences, which both the public and private sector have to comply with. The Dutch government continued to support participatory processes with a declaration in 1998. Emphasising the need for better integration of residents in policy making, it motivated public participation far beyond legally binding procedures. The Ministry of the Interior commissioned the Dutch Centre for Political Participation IPP to investigate existing forms of participation throughout the country. The collected experiences of this study were published in 2002, setting a benchmark for further development. It marks a shift in the perception of the participatory process from a legal procedure to a self-evident planning instrument.

According to the different natures of the individual projects, the Artery partners have used a variety of public participation methods. The young people's panel in Speke and Garston and the participatory appraisal

Young people's panel discussing the future of the Speke and Garston Coastal Reserve

exercises in Stockport, offer an insight into the experiences along the river Mersey. In Stuttgart the river Neckar was subject to a design competition. Experiences have been shared from the communal brainstorm sessions at the Baai van Krimpen and Loswal III project, as well as the public initiative behind the redevelopment of the old Vuyk shipyard and the windmill Windlust. Extensive consultation with target groups like the youth and disabled in the Ruhr Valley, as well as public co-operation in operating the River Bath Ruhr, will serve as examples in this chapter.

Basic Characteristics of Public Participation

Public participation in spatial planning differs greatly from the formalised procedures of the past. It characterises an informal and open process and serves as an effective planning tool, in addition to standardised legal procedures. The general definition and the undetermined character of public participation, allows for a variety of different forms of participatory processes. For this reason it is obviously difficult to generate a standard procedure, nevertheless a basic guide to the principal elements can be established. Public participation is the interaction of all three core elements of a resident – project relationship. According to the degree of public involvement in the spectrum of resident – project relations can be characterised by the following aspects and their correlating effects:

Information   Consultation   Participation

It is essential to recognise the fundamental differences between each of these terms.

Information

is the foundation of all participatory processes. Usually initiated by project planners, it is a one-way form of communication which distributes information to the people affected. Thorough information provides the public with a detailed insight into plans and their implementation. It raises awareness, and is the first step towards further participation. This stage of public involvement is characterised by → Co-knowing.

Consultation

describes the stage where the project team or administrative institutions consult people affected by a project. Information and specific knowledge is gathered from local residents and their ideas and concerns are acknowledged. It is a fruitful two-way relationship between project managers and residents. Through the implementation of a project, this aspect offers both the public, as well as the developers the opportunity for → Co-thinking.

Information is the root of all public participation, and an important preparatory step in the process. The common challenge throughout the Artery projects is the regeneration of river landscapes. A lack of knowledge about the potential of public space combined with difficult access, has led to negligence and desolation in many areas. To counter these developments all projects rely on public acceptance, which can only be generated through knowledge. Supplying the public with information draws their attention to the development of a site. A strong emphasis on maintaining reliable sources of information can therefore be found throughout the Artery pilots. The British partners produced regular update reports from their reference group meetings, which were sent out to participants and made available to a wider public online. An extensive website also provided information on the regional context of the Mersey Basin Campaign. The Dutch partners issued a printed bulletin every month as a supplement in local newspapers. This kept the interested public up-to-date with the general redevelopment scheme for the Hollandsche IJssel, and also on specific Artery pilot projects. Information creates awareness, and was therefore used extensively in the Rhine-Neckar region to make people aware of the different parts of the Agendapark Living Neckar. This facilitates participation – leading to increased acceptance and adding to the sustainability of regenerated public spaces.

The next element of the participatory process is public consultation. Once the information has drawn attention, and in its wake people, to a developing site, it is necessary to meet expectations. Consultation enables planners to map desires and needs that exist in a region, tap into the knowledge of local residents and use this in their planning. Where expectations are met people are more likely to accept a redeveloped site as their own public space. The sustainability of a project increases when more people become involved and a broad feeling of ownership has been generated. This aspect of public participation is vital to projects that have to consider the users' special needs. An essential part of the large-scale redevelopment scheme in the Ruhr Valley is to draw new visitors to the river. In a drastically changing economic environment, development programmes need to take into account shifts in society. The Witten Labour and Employment Company's (WABE) pilot for example, focuses on groups with impaired mobility. It offers easily accessible tourist facilities and daytime recreation for the physically challenged and also financially underprivileged people. Representatives were invited to share their knowledge to help increase disabled access, while at the same time creating jobs to reintegrate long-term unemployed into the workforce. Their respective needs

Public hearing at the Municipality of Capelle: residents were informed about the general idea of the planned development

Participation

integrates the public in the planning of a project. The advice given by the public determines and improves the overall goals. This active participation creates an advanced two way working relationship between the general public and planning bodies. As it minimises potential conflicts, it is likely to lead to a reduction of the long-term costs of implementation. This dynamic process can lead to shared decision making where project planners share responsibility with previously identified stakeholders. It supports the self-determination of the public and generates ownership of a site. At this stage public participation turns a project into a joint effort. The public is → Co-operating.

made consultation with the targeted groups indispensable. Only projects that meet these special demands are accepted and viable on a long-term basis.

Co-operation between regional planners or project developers and residents is an essential part of public participation. When information sparks public interest and consultation identifies residents' intrinsic knowledge, it can be tapped into as a valuable resource for further development. The commitment of local clubs and residents to the operating society for the natural pool on the banks of the Ruhr in Wetter, represents this stage of public participation.

As resources are needed to sustain public spaces, as well as public facilities like swimming pools, pro-

The pilot partners of the Vuykyard meet to discuss the outcome of public consultation // In small groups citizens participate actively in the decision making process in Capelle aan den IJssel

jects often face difficulties in their financing. This strain can be alleviated when the public increases its share of responsibility for its own environment. However, co-operation only functions if project planners are prepared to open up the planning procedure. At the Ruhr the original plans to expand an existing leisure centre by vast open-air natural swimming pools, were put on hold when the operator's economic situation ruled out further investments. The Regional Association Ruhr (RVR) with the Ennepe-Ruhr County and the municipalities of Bochum and Witten are partners in this joint venture and started to look for a new location for this natural pool project. Meanwhile the municipality of Wetter publicly discussed the future of the existing open-air pool and one of the options included the conversion into a natural pool. Interested residents gathered the support of 4,500 local people through a petition, which eventually resulted in the foundation of

Degrees of Participation

The right degree of public participation has to be assessed for every new project individually and the experiences of the Artery pilots are very encouraging.

a new operating society. Together with the RVR and the municipality of Wetter it co-operated in the realisation of a natural pool. The new partner took over the operation and maintenance of the pool in a shared financial partnership.

Approaching Public Participation

Successful public participation requires appropriate preparation and a project that allows for an undeter-

mined and open planning process. Responsibilities need to be made clear from the start and expectation management is a key issue in preventing frustration. Before any participatory process can start, it is essential to thoroughly assess the implications of a project. The need for public participation has to be taken into consideration, as well as its scale in terms of the amount of time, money and people involved. The collected experience of the Artery projects provides a guide that will help identify the appropriate tools for a successful public participation process.

Identifying Possible Partners for Public Participation

Public participation is a process often associated with the efforts of a government to involve its citizens in policy making. But this view limits public participation unnecessarily. The process can also be initiated by public bodies, public-private organisations, by companies, societies, platforms, associations and clubs. These different initiators can be identified by their respective position within the resident – project relationship. Different types of relationship need different kinds of processes.

Planners Seeking Assistance

As many projects of public concern are still initiated by governmental institutions, most participatory processes are managed by local governments too. Public participation is an increasingly important part of local policy. For example, for the pilot Baai van Krimpen and Loswal III, the Projectteam Hollandsche IJssel encouraged and supported the municipalities of Krimpen and Capelle aan den IJssel to involve the public from the very beginning, even though they were not obliged to by law. In this case the participatory process was initiated even before the first sketches were drawn. In public brainstorming sessions residents put their ideas forward and contributed to an open plan process. A professional landscape architect then turned these ideas into alternative proposals for the redevelopment of the site. The municipalities in turn checked the proposals for their technical viability. The resulting plans were surprisingly simple. The proposed changes were minuscule as the residents appreciated

---

In Stockport (left) and in Capelle (right) detailed plans and showcases were used to facilitate discussion between experts and residents

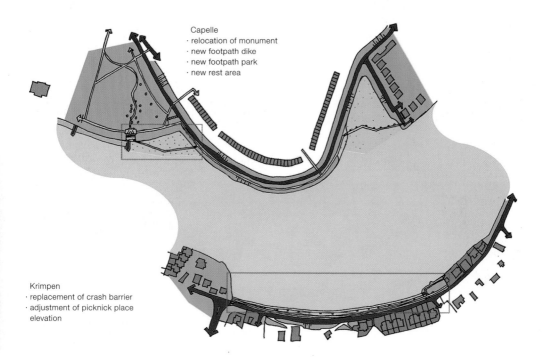

Capelle
· relocation of monument
· new footpath dike
· new footpath park
· new rest area

Krimpen
· replacement of crash barrier
· adjustment of picknick place
  elevation

their unobstructed view across the river bend. The active involvement of the residents increased public awareness for the cause of the Hollandsche IJssel and a growing sense of ownership resulted in a more cost efficient implementation of the regeneration scheme.

Public participation can also be initiated by semi-governmental institutions. These institutions are often joint ventures of the local government and private businesses, and therefore have a high emphasis on the efficiency of the financial means used. The Mersey Basin Campaign, an initiative backed by the government and partly funded by private businesses, used public participation to assess the local relevance of specific details of their plans. The neglected and run down riverbanks at Stockport were in dire need of regeneration and the threat of continuous vandalism demanded considerate planning. In this instance, public participation served as a tool to increase the efficient use of available resources, so that unnecessary features could be identified and avoided. Participatory appraisal, described later in this chapter, was used as a method to assess the future social impact of the Mersey Vale Nature Park, to increase

future acceptance, and therefore minimise neglect - the main cause of vandalism.

Residents Seeking Access

A public initiative for a participatory process, as in the case of the Vuykyard and the Ruhrbath pilots, is still an exception. During the reconstruction of the Vuyk shipyard, the concerned public asked the municipality for access to the planning procedures, as they feared the opportunity for a green, recreational area would be lost to housing. With their previous experience of running a neighbourhood council, they established the Petit Comité Vuyk specifically to monitor developments around the former shipyard. This committee was only made up of a small group of people, but their meetings remained public and were advertised, so that other interested residents could join in. Although not a legal part in the decision making process, the municipality referred to the Petit Comité Vuyk as a reference group, which closely followed the

The plans for redevelopment of the Baai van Krimpen changed in the course of the process resulting in a modest but fitfull design

redevelopment of the derelict site. In this way the public gained access to the decision making process, and participated in the planning procedure. As an integral part of the organisational body, it aimed at completely changing the plans for the area. The housing plans were eventually dismissed allowing for public space, which was greatly appreciated by the locals.

An unusual form of public participation occurs when private companies try to involve the public in a planning procedure. This only happens with companies that already have a strong social interest. It is usually done to better assess their cause and lobby it amongst politicians and public alike. Although the previously mentioned Witten Labour and Employment Company (WABE) is not a privately owned company, it nevertheless serves as an example of how participation works in the private sector. WABE was an initiative set up by the local government and the Ennepe-Ruhr County, its shareholders included various other groups and social institutions. It is a non-profit company aimed at the reintegration of long-term unemployed people into the work force. As WABE is required to cover its own operational costs, it has a profound interest in participatory processes. A better understanding of the needs of future users helps to improve accessible facilities on their site. Additionally it also ensures better business, as more visitors are likely to come to the site when these objectives have been met. Public participation functions as a tool to assess economic viability and helps WABE to fulfil their requirements of running profitably. Their target group approach will be explained later in this chapter.

Regardless of the initial approach to public participation, a successful process must be initiated as early as possible in the planning procedure. A good way to start the participatory process is through a stakeholder analysis, as described in the next section.

Determining Potential Partners

Different interests need to be considered throughout the course of any project. The important decision that has to be made at the very beginning is who to involve in the planning procedure.

The Neckar is back in the focus of the local people

Stakeholder

Stakeholders are people, groups or institutions, whose interests are affected by a policy, programme or project, or whose activities could affect the project either positively or negatively. They possess information, resources or expertise needed for implementation.

A stakeholder analysis helps to identify suitable partners for a project and assesses the social environment in which it operates. It characterises those involved by function, separating decision makers, users, implementers and experts. The results of a stakeholder analysis determine the appropriate type of participation, keeping in mind that the role of the same stakeholder can change at successive stages.

A stakeholder analysis is a process with three characteristic stages:

Identify   **Determine**   **Involve**

The stakeholder analysis prevents the risk of overlooking important players. To ensure the practicality of public participation it is sometimes necessary to make a choice about who to involve in the decision making process. Not all stakeholders will be able to take part, but credibility necessitates transparency about the criteria of the choices made.

Therefore, it is important to identify the different stakeholders prior to the project and to compile information on their background during the process. This enables a project team to assess if the chosen stakeholders mirror the community's overall structure, and where necessary to extend the consulted group accordingly. In general a process of public participation should reach an average ten percent of the affected local population in order to achieve a representative result. One way to gather this information is participatory appraisal, the public participation exercise used extensively at Mersey Vale Nature Park.

Public Participation

Inviting Partners

In general there are only two possible ways to contact the public: either invite people to talk to you, or go to them and ask questions. Stakeholders differ in their relationship to a project, making appropriate communication necessary. Their distinct internal structure offers a guideline. An organisation of disabled people for example, is a thematically related organised target group. Children or tourists are referred to as specific non-organised societal groups, and non-organised individual citizens include neighbours and occasional users. Communication approaches reflect these characteristics. Personalised invitations increase the impact among representatives of organised stakeholders, while local newspaper ads work better to integrate non-organised stakeholders. Both communication channels obviously differ in reach.

In the Artery pilots several approaches to potential stakeholders have been used. Special effort was taken to integrate difficult to reach groups by contacting them in co-operation with local partners with already established access. Sports clubs for example, are a typical venue for young adults. In Speke-Garston members of the sailing club were involved in the stakeholder forum. As users of the site they showed a keen interest in developing the Coastal Reserve.

Other potential stakeholders can be reached through schools, which can also be approached to help find appropriate methods to consult children. At the Stockport Mersey Vale Nature Park this resulted in a dedicated survey among pupils aged 14 to 17, a group of frequent users of the site who otherwise might have slipped through the consultation process.

The run down pool in Wetter in dire need for renovation

Success Story: A Friendly Takeover In Wetter, the proposed closure of the open-air pool on the banks of the river Ruhr sparked an initiative amongst users of the facility.

Local people established a foundation and a deal was made with the municipality to take over the bath as an operating association. A growing awareness of environmental and social causes, as well as their own recreational needs, resulted in increased self-determination. Participation in decision making led to a share in responsibility. 'Unser Freibad am See e.V.' (Our Swimming Pool at the Lake Association) is now responsible for the daily running and economic viability of the natural bath. The contract with the municipality of Wetter requires the association to reduce the operational costs of the river bath. They in turn initiated a participatory process to involve the general public in this ambitious venture. Public participation aligns economic success with benefits for society – a common theme throughout Artery's projects.

Stakeholder Analysis

A few steps to consider on the way to a reliable stakeholder analysis:

~ Define the key issue and indicate the stage that will be subject to a public participation process; putting it in question form helps.

~ Ask obvious stakeholders (the project team or those who will be part of the participatory process, as they initiated it) to brainstorm for other possible partners from as many perspectives as possible; note everything, don't judge.

~ Go from general into detail, starting from groups working towards individuals, before collecting contact info.

~ Check, sort by function and degree of involvement and look for big gaps, identify interests, conflicts and existing relations with or amongst stakeholders. Guiding questions are: what is expected, where is the benefit for each stakeholder, what resources can be committed, where are conflicting interests?

~ Use the results and contact stakeholders; inform them of their role (management of expectations).

Groups of difficult to reach Stakeholders

Previous participatory processes have shown that stakeholders are insufficiently represented if they belong to the following societal groups:

- ~ Small children - through lack of appropriate communication

- ~ Youth aged 13-18 – lack of appropriate approach

- ~ Men aged 25-50 - lack of time and interest

- ~ Senior citizens over 70 - often through lack of general meeting points or mobility

- ~ Ethnic minorities - due to language barriers

- ~ Users living further away - as they are not part of the local community

Experience in Artery projects has shown that these groups are difficult to reach for developers, because of either the structure of approach or inaccessibility to the project. Special attention has to be given to the design of a participatory process, in order to reach out to these groups.

The young people were invited to a walk about on the site, and were then asked to imagine themselves as bird watchers, bikers or employees on a lunch break in order to collect ideas from different perspectives. The pupils then gathered again to prioritise their ideas and to give a summarising verdict of the situation.

In order to reach out to migrants or ethnic minorities it is advisable to distribute information through the respective community centres, and if necessary ensure that it is available in different languages to avoid obstacles.

Special events held on a site itself, combined with extensive press coverage, help to reach out to non-organised individual residents. At the Vuykyard in Capelle aan den IJssel for example, a public barbecue was held to draw people and their attention to the project. The high turn-out of interested people encouraged the planners in their participatory approach, and also lowered the barriers for people

In Stockport public participation was taken to the street motivating people to take part in the decision-making // The project team of the Mersey Basin Campaign planning the different stages of public participation

Expectation Management

Public participation thrives on the involvement of people. However, enthusiasm can lead to disproportionate expectations, diminishing the results of a participatory process. Clear communication about the role of public participation within a project ensures that a realistic perspective is maintained throughout the process. Open discussion points need to be distinguishable from given facts that are unalterable, due to technical, financial or legal restraints. Management of expectations prevents disappointment and frustration and is a corner stone of successful public participation!

who wanted to take part in the process. The potential site users contacted directly at the event, might otherwise have slipped through the consultation process. Although the meetings of the Petit Comité were open to everyone, their well established and ingrained decision making process might have formed an unwanted barrier. The event successfully broadened the base for the participatory process and with its high acceptance, secured ownership of this pilot among residents.

Involving the Public in Decisions
Collecting Information

Referring back to our initial sketch of the different aspects of a participatory process, an essential part of public consultation is the collection of information. Once the initiator of a project has identified the potential stakeholders, it is then necessary to gather the right data.

Participatory appraisal

The technique of community consultation is especially useful when trying to involve a wide group of people in a project. With its easy and playful approach it focuses on going out to the people. Line Tools and Prioritising Circles allow people to simultaneously express, evaluate and contextualise their ideas providing project planners with qualified information.

The grand opening of the Vuykyard in
Capelle aan den IJssel

Success Story: Community Park Vuykyard A group of committed residents in Capelle aan
den IJssel and wide-spread concern throughout the community, started a campaign for participation. The Petit Comité distributed 300 questionnaires in the local shopping centre asking
people to comment on suggestions for the future use of the derelict shipyard, and to prioritise
issues according to their own interests. The pattern of answers provided qualified information for
a detailed map of the potential development. The results showed a majority in favour of unobstructed access to the Hollandsche IJssel, a playground for children and drastically reduced
housing to retain the open character of the site. This survey proved a valuable database in con-
vincing politicians to change the plans for the former shipyard. It was also used in discussions
about the design of the park, as new proposals could easily be measured against the qualified
results. At one stage they even invited the former owner of the shipyard to become a member of
the Petit Comité, as his knowledge of the terrain was unsurpassed. He proved to be a very valuable member, especially for the design of the redevelopment plans.

Participatory Appraisal

The Mersey Basin Campaign used participatory
appraisal as an instrument to directly approach the
public during the planning of the Mersey Vale Nature
Park. The problems along the Mersey bank were that
people felt unsafe and were appalled by the derelict
site in Stockport, which prevented them from seeing
the land as a potential community park. The project
team kept this in mind as they questioned visitors to
gain a better understanding of what the public wanted
from the redevelopment of the area into a community
park and nature conservation zone.

Once the potential stakeholders have been identified, it is advisable to find out where they meet and
carry out the first steps of consultation there, in order
to gather information about interests from people
who work and live in the area. The experiences of
the developers of the Mersey Vale Nature Park at

Stockport show that this can take place at a variety
of locations. The initiators of the participatory process
went out to schools, sports clubs, church groups,
shops, libraries and business parks in the vicinity.
They also frequented the site itself. At the end a total
of 23 of these consultation exercises were carried
out. This removes the boundaries that can prevent
people from participating in a decision making process. It promotes the project among people who were
previously unaware of the potential of the site and who
would normally not come to a local meeting or public
hearing. It should be quick, simple and fun to use in
order to attract as many people as possible.

The process of participatory appraisal starts off
with general, non-guiding questions to determine the

At Stockport special efforts were undertaken to reach out to young
children

Public Participation

What are the three biggest problems with the riverside land in Heaton Mersey?

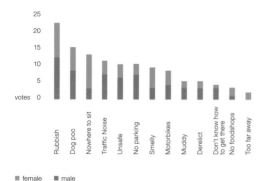

■ female ■ male
(total people asked=48)

What are the top three things you would like to see next to the river in Heaton Mersey?

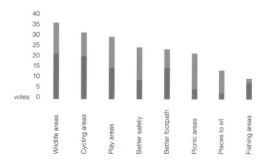

■ female ■ male
(total people asked=79)

The extensive consultation in Heaton Mersey helped to prioritise the measures to be taken // Participatory Appraisal at Stockport Library // Bright colours and simple codes make participatory appraisal easily accessible while collecting detailed information

main issues at hand. The participants were asked to note down their answers to these core questions on post-its and attach them to flip-charts. These charts were prepared with so-called Line Tools, a line ranging from yes to no, or Prioritising Circles, distinguishing between important and marginal issues, allowing for quantifiable and qualifiable results.

Participants attached an easy to use sticker key to each of their remarks. A system of codes indicated their sex, age, ethnicity and location of residence or work, as well as whether they were dog owners or had been asked before - to avoid double counting. The information could then be checked against data obtained from the last census. This cross-check of the extensive participant information enables the developer to assess if a representative sample group from the local community has been consulted. The results clearly showed the conflicting interests of

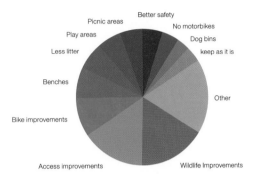

What People want in the Riverside Land

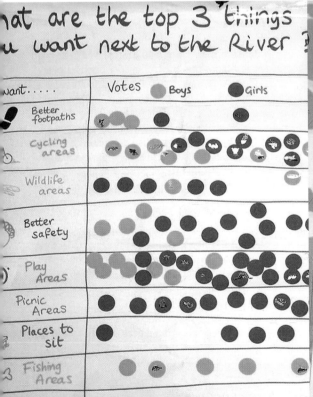

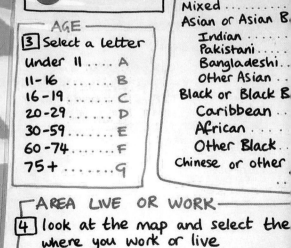

action bound pastimes like motor-cross and mountain biking, with seekers of tranquillity, such as bird watchers and walkers. The participatory process continued as the results were presented to a reference group for further discussion. Based on the collected data the design of the park could ensure that there was a clear separation of the different and sometimes conflicting interests. As the stakeholders' needs could be assessed more reliably, public appraisal increased the residents' awareness of the potential of the riverside, which added to a growing sense of ownership for the site in the community.

Competitions

Another efficient tool to collect ideas from the public, is a competition. The main obstacle in regenerating the Neckar between Wernau and Altbach is the narrow river vale. The river shares this limited space with a rail-road and a busy federal highway, as well as extended industrial sites on its banks. Careful consideration of potential conflicts needed to go into the redesign to enhance the precious stretch of land

in this dense environment. The Stuttgart-Neckar pilot decided to hold a photo-competition, open to everyone. People were asked to submit two pictures, one showing their favourite spot on the site and another one depicting the part they liked least. The pictures, submitted by a broad public, helped to map the site from a user's perspective. They identified the most common uses of the site, and drew immediate attention to the most urgent problems. The results were discussed with the residents and an agenda group was set up to give further advice during implementation. This is of general importance, as participation can only be successful when active measures are taken to turn the public's ideas into reality. Only then can a lasting acceptance be generated and the sustainability of a public space increased.

Student Competition

The photo competition was complemented by a transnational design competition, another tool to

Students participate in a design competition for the Wernau Neckar-River Reserve

obtain creative input for a project. It was carried out amongst European students of architecture. In co-operation with the Universities of Stuttgart, Dortmund and Karlsruhe, as well as Rotterdam, their partner university in the Netherlands, students had the chance to work out a master plan for the area concerned. The fresh view on a difficult situation added challenging ideas to the creative process. While the young professionals could test their skills, the project benefited from the unconventional approaches proposed in their work.

Public Brainstorming Sessions

At the Dutch project of Baai van Krimpen and Loswal III, the general public were invited to open brainstorming sessions, so-called 'Meedenksessies'. The challenge was to find an appropriate design for the regeneration of a heavily contaminated stretch of riverbed. The sanitation of the Hollandsche IJssel offered the opportunity to simultaneously redevelop the riverbanks. The location possesses both characteristics of the Hollandsche IJssel, the open Polder landscape on one side, as well as the densely populated riverbank on the other. These differences correlate with the conflict between natural preservation and recreational needs. Public participation was used as a way to find a fitting solution to this challenge. The unique aspect of this participatory process was that the public was not limited to certain groups or stakeholders. People living in immediate proximity to the site were approached by the municipality in personalised mailings. Others were invited through articles in local newspapers. It's interesting to note that during the first meetings, that were held separately on the two opposing riverbanks in Capelle aan den IJssel and Krimpen aan den IJssel, the public demanded the involvement of the respective residents across the river, as they would be the ones looking out on the regenerated riversides.

These public brainstorming sessions changed in character through the course of the process; from information and consultation evenings with a reference group, to a working conference. At the very creative beginning of the project the meetings were

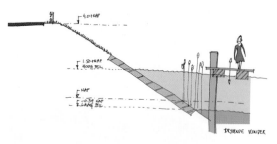
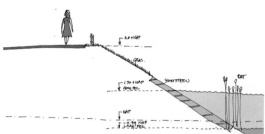

frequent, which allowed for the collection of as many ideas as possible using the methods of participatory appraisal - such as the line tools and prioritising circles. A landscape architect condensed the collected ideas into three principal sketches. Once a basic concept had been agreed on, the meetings became less frequent and turned into working conferences. In

Expert-Layman-Dilemma

A severe obstacle is the dreaded Expert-Layman-Dilemma. It describes misunderstandings between professional experts and the participating layman public. The first public brainstorming sessions at Baai van Krimpen and Loswal III experienced some difficulties of this sort, as misunderstandings between experts and public occurred, as well as discussions among the experts themselves. In most cases it is a problem concerning technical terminology, or a clash of a systematic versus an unsystematic approach within a work group consisting of experts and public.

this second round, the participants were able to comment and alter the plans under guidance of a panel of experts who offered support on technical details and gave an outline of implied limitations. The design chosen for realisation was surprisingly modest. The open view over the river was highly valued, which ruled out extensive planting. The final proposal offered very few recreational features, as neighbours feared these would attract too many people to the already limited space. Public participation helped to prevent investing in unnecessary features, while drawing the government's attention to previously unnoticed problems like inadequate road safety.

The trustworthiness of this form of public participation depends on the credibility of the experts involved in the process. The best results are achieved when discussions amongst experts on the panel can be prevented. Otherwise these discussions confuse the participants and severely disturb the process. An independent presenter has proved helpful, as it forces everyone to use viable arguments. Someone who is otherwise not involved in the implementation of the project can better mediate between the public and the expert panel.

The Target Group Approach

Another way to collect valuable information and advice for a project is the direct approach of target groups. When the stakeholder analysis delivers specific social groups as people with a strong interest in a site, these can be directly consulted by means of interviews or people panels.

Different visualisations facilitate discussions of various options at Baai van Krimpen

The young peoples' panel inspects
the future Coastal Reserve

Success Story: Active Involvement counters Vandalism In Speke and Garston young people were asked to participate in the development of the Coastal Reserve. The area suffered severely from vandalism and had turned into a 'no-go' area. A thorough participatory process to assess public needs became all the more important. The stakeholder analysis identified young people as an important target group. Meeting the demands of young people, public participation could prove a useful tool for improving sustainability and potentially reducing vandalism, as it generates ownership of the site. The potential of a participatory process determined the further involvement of young people in the project. Using professional interview techniques, a social research company asked local pupils how they thought the site along the Mersey should be developed. Organised as a people's panel, the pupils were comparable with a jury. Assumptions or ideas from earlier rounds of public participation were subject to the criticism of this group. 16 young people were asked to adopt the position of other stakeholders, and use these different interests as a background to evaluate a set of different propositions against. The process furthered the pupils' own understanding of the complex demands on the site, while also ensuring that the project could be adapted to better meet the young people's interests. New plans for the site now included a noisy area, where kids could meet on their own terms without interference from other users. The combination of enhanced facilities and better maintenance, as well as increased awareness and distinct areas for individual needs, benefits the sustainability of this public space.

At the WABE pilot of accessible tourist facilities for example, two special target groups were identified besides general day trippers: young people and disabled people. Both user groups were approached through their respective organisations. Youth Parliaments and disabled representatives have built an extensive network in the Ruhr area connecting local initiatives to a regional context. Representatives of these organisations participated in round table meetings to add new ideas to the project's implementation and development. Special events were organised, like a bicycle-rally, to raise awareness for accessible bicycle routes. Ideas included in the planning, such

as an open camp-ground for youngsters, resulted in the promotion of the pilot amongst the target groups, and increased the feeling of ownership for the newly developed accessible tourist facilities.

Keeping the Public involved

The above mentioned techniques produce a variety of different ideas. The next logical step is to implement these ideas, as well as some kind of control over further processes. Public participation is structured in separate stages, in order to deal with results, and to involve the public in the implementation process. The use of the appropriate communication tools is essen-

tial to keep participants involved in the interaction. The different stages within the participatory process require different usage of available communications: co-knowing works by advising media through presentations or articles in the local press, and distributing information among a broad public. Co-thinking on the other hand, requires a communication, such as the participatory appraisal process, interviews or discussion groups in order to draw on people's ideas. Co-operating needs provide yet another setting for succesful interaction. It requires small groups with limited hierarchy and similar knowledge of the matter, to create an open work atmosphere with direct communication.

In Stockport the biggest project challenge was overcoming the derelict character of the inaccessible and surprisingly unknown site. An extensive public participation process was initiated focussing on how to reach out to as many people as possible to raise awareness for the public space; this took 18 months in its entirety. The process started with a leaflet dis-tributed throughout the local community containing information on the project and a brief consultation exercise. A total of 103 potential stakeholder groups were contacted and asked to share their opinion. Participatory appraisal was used in 23 consultation exercises on the site and in the community, raising interest among residents and adding to the project consultation database. The objective of extensive reach was met by establishing a strong feeling of ownership within the community.

Crucial to the success of a participatory process is the implementation of the resulting ideas. Therefore, the follow-up process needed to be streamlined, in order to allow discussion in greater detail. 18 stakeholder groups with an on-going interest were invited to participate in a reference group. It included orga-nised stakeholders, as well as individual residents and served as a forum for feedback from the community consultation. Its purpose was to engage key stakeholders in the project for provision of expertise. Only then

could the project really benefit. The image of a public space can be altered as acceptance grows through recognition and realisation of the public's ideas.

Public participation at the Vukyard pilot came from within the community, and it required an organisational form to facilitate further procedures. The Petit Comité Vuyk was created by interested participants at the first stage of the open plan process. With this reference group, co-operation on the Vuykyard could go ahead in detail and proved to be a reliable entity within the work process. Together with the civil engineers, they were responsible for the design of the Vuykyard, and also ensured the further involvement of the general public by making presentations at information evenings.

Organised stakeholder groups often have specific knowledge relevant to a project. At the Stuttgart-Neckar pilot, work groups were chosen to integrate this knowledge and to deepen the participatory process. Both the nature organisation NABU and the local fishing club participated in the project, successfully co-operating in its planning and implementation.

Benefits and Difficulties

Public participation takes an effort. It costs time and money for all parties involved. A thorough consultation produces an overflow of information and ideas. Although the quantity of data is highly valuable, it can be a burden to process it, and sometimes elaborate participation results in only a few changes to a site.

The Artery projects show that the alleged weaknesses of public participation can be turned into strengths. The simplified yet tailor-made proposal for the new riverbanks at Baai van Krimpen and Loswal III saved the municipalities from unnecessary investment in undesired features. Any project can gain

Obstacles were clearly identified by participating groups of a rally for accessible bicycle routes in the Ruhr Valley // Reference groups monitor the participatory process through its stages at Stockport

Public Participation

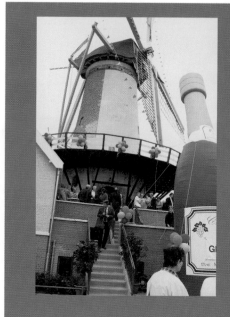

Success Story: Participation promotes Commitment The reconstructed windmill Windlust on the Hollandsche IJssel is an example of how the traditional responsibility assignment of institutional implementation and public consultation can be successfully turned around. Motivated residents, concerned about the state of the region's cultural heritage, took action and initiatives to save a historic windmill. To realise this ambitious project they initiated a participatory process turning the usual top down approach into a bottom up initiative. The involvement was not aimed at the general public but at local governmental institutions to lobby their cause on a larger scale. Public participation leads to residents who will eventually take over the initiative for a project. And when they find support for their cause, the long-term goal of increased public self-determination and agency is realised.

from public participation. With deliberate organisational integration of public participation, the extensive information gathered in the pilots along the Mersey, proved to be more than beneficial for the respective sites. More suitable and sustainable solutions can be developed, as the knowledge of people living in the area is integrated.

With good management public participation can prove to be a time-saving tool over the course of a project, outweighing the costs and efforts incurred at its initiation. People are less likely to block projects when they are asked about their concerns right from the start. This can also save money, especially in big projects where the costs of a participatory process are marginal compared to the potential damage legal procedures can incur. A welcome spin-off is the improved image for both project and initiators in public opinion, as its verdict is already an integral part of the process. The better understanding of the needs of the public leads to the development of a more appropriate, and therefore better policy. Projects maintain

a higher sustainability with simultaneously lowered maintenance costs, as people are encouraged to take ownership. The overwhelming turn out at the opening festivities of the redeveloped Vuykyard in Capelle aan den IJssel gives proof of people's acceptance and enthusiasm for their public space.

A successful public participation process can help to achieve general improvements within a community, too. Businesses, administrative institutions and politics can improve their relationship with residents, as public participation can help de-escalate existing conflicts. The emphasis on equality in communication enhances a community's social, human and ecological capital. Public participation draws on people's interests and concerns in the planning processes, and facilitates bottom up initiatives by people taking responsibility for an improved environment. The experiences individual people are likely to have during this process of decision making promote their autonomy and self-determination in problem solving and provide motivation for future initiatives.

For a successful public participation process the following aspects need to be considered:

~ Provide the public with detailed information on the planning of the intended project.

~ Collect ideas and concerns from local people through consultation.

~ Motivate residents to participate in the planning and – where possible – the realisation of the project.

~ Be sure to identify all possible stakeholders in a stakeholder analysis.

~ Discuss and determine the role of potential partners in a project and involve the public accordingly.

~ Use continuous information to further stake-holders' qualifications and develop the project team's human resources.

~ Clearly define the project's financial and technical temporal boundaries as well as its time frame.

~ Manage the public's expectations over the outcome of the participatory process.

~ Keep a comprehensible and – where possible flexible schedule

~ Use the results of public participation!

Public participation builds and maintains interest and commitment in environmental and social issues. It should therefore be a priority from the very beginning of a project.

Text: Klaus Mandel

Public Awareness

Adjacent to the new town entrance at Ladenburg, children can now play and experience the Neckar // Pupils in Stockport helping to design the artwork for the entrance of the Mersey Vale Nature Park

Introduction

Over the last few decades sustainability has become a central issue across Europe. The United Nations and European Union governments acknowledge that economic security, social justice and ecological balance are interdependent key aspects of sustainability. The necessity of actively involving the general public in decision making processes (public participation) has also been recognised. It is a vital element for future regional and global development.

A major precondition for public commitment is awareness. In order to enable people to participate they need to have the knowledge and experience that will build the basis for rational and appropriate choices. To create long-term awareness, it is essential to provide the public with new information that affects them on an emotional level. In order to achieve lasting changes in behaviour and active public participation (the ultimate goal of raising awareness), public perception and attitudes have to be affected.

Under the motto "Think global; act local", it is important to educate the public on a regional level. As well as providing factual information, educating the public also increases their attachment to key regional features. People can only start to identify with and take responsibility for their environment, when they become aware of what's in their local vicinity. Raising public awareness is therefore an important step towards a sustainable environment, as only regional identification can lead to a sense of responsibility and ultimately, active public commitment.

For the Artery partners the strategy of raising public awareness was essential. It was important to increase overall public consciousness for regional and environmental development. For regional development planners it is important that the local population values their surroundings. The relationship between the public, their cultural heritage and natural regional elements, such as rivers, must be enhanced in order to promote awareness.

By inducing a public change of mind the partners pursued two goals. Firstly, many local people felt indifferent towards their local riverside. Through an awareness campaign the Artery partners hoped to change this neutral attitude to a sense of value, creating active commitment. While most people felt either indifferent or positive about their environment, a minority behaved destructively. Vandalism demonstrated these feelings. Changing negative attitudes into respect was therefore the second goal the Artery partners hoped to achieve. These were the educational tasks facing the partners. Therefore, raising public awareness contributed to the projects' sustainability.

In order to draw people's attention to general key issues or particular regional activities, various tools can be used. Examples from the Artery pilots showcase which methods proved to be fruitful in raising public awareness.

What is Public Awareness?

Awareness means becoming conscious and mindful of a certain issue. It implies knowledge gained through one's own perspective, through formal and non-formal learning and experience. Awareness can be raised externally through the actions or attitudes of others. It puts an individual in the right position to make decisions and is therefore a basis for action.

Raising public awareness is the process of informing the general public and increasing their level of consciousness about a certain topic. It usually includes giving advice on how to act in a specific situation. It also involves presenting it in a way that is interesting and entertaining. Events and exhibitions are popular methods of raising awareness for a certain cause. The next step, and an integral part of the awareness raising process, is to increase the attribution of responsibilities to the individual.

Regional development programmes involve raising awareness for particular projects or general regional and environmental issues, and usually contain educational elements.

Individual development projects increase interest, encourage acceptance and raise the value of implemented actions. The region increases its attractiveness for residents, which leads to stronger regional identification and active public participation. Public awareness for the region often goes hand in hand with environmental awareness.

Environmental awareness is the recognition of environmental concerns and values. Economic, social, professional, cultural and educational factors play a part. These aspects have to be taken into account when attempting to strengthen environmental awareness.

The Importance of Raising Public Awareness
Regional Awareness

Many European regions have been affected by the structural change from industrial to service socie-

A child playing at the Neckar. The Neighbourhood Association Heidelberg-Mannheim puts emphasis on these experiences that increase children's regional and environmental awareness // Guided canoe trips on the Neckar offered children and adults valuable new sights on the river and its banks

ties. Regeneration of industrial brownfields is a task that these regions will be dealing with well into the 21st Century. When working towards a sustainable future, regional development planners are faced with the challenge of bringing the region closer to the population.

In order to appreciate a regenerated river landscape, people have to first see its value. Which means bringing them back to 'their' river, to understand how a healthy river landscape can contribute to their own living quality. People have to get in touch with their cultural heritage, in order to be able to understand its meaning.

Awareness for a region's identity and therefore for the region's main features, creates a public regional identity. This affinity helps people become active. The population's regional awareness often results in active public participation. With the knowledge and experience gained through public awareness campaigns, they are able to engage in decision making processes within their community, can develop their own opinions and become actively involved. They stand up for what they think is good for their region and for the developments they would like to see.

Another consequence of post-industrial restructuring is greater competition among regions to attract businesses and qualified labour. In the face of such competition it is important to stop the local population from moving elsewhere, and to establish incentives for people from other areas to move to the region. Promoting regional identity, and thereby strengthening the population's identification with the region, helps prevent people from leaving, and also attracts new people with new input. People, who are aware of their region, develop a bond and are less likely to relocate.

Raising regional awareness is also important, in order to gain broader support for individual development projects. In times of local authority cutbacks the

public is often critical towards investments in which they cannot see an immediate benefit. It is therefore necessary to systematically approach the public and increase their level of knowledge about the implemented or planned developments.

Environmental Awareness

Public awareness for environmental concerns started to emerge in Europe in the 1960s. The transition from an industrial to post-industrial era was setting-in across many regions at the same time. The social and ecological effects of centuries of exploitation and alteration of nature became evident. Major industrial developments, which often led to spatial separation from their natural environment, meant the public became disconnected from important features of their local habitat. Dying forests and polluted rivers unable to sustain fish were further examples, which demonstrated the urgency to act.

Individual European countries started counteracting these effects to different extents and with differing methods. It was a slow process, mostly dominated by new legislation enforced by governments without the involvement of the local population.

Over the last forty years it has become evident that sustainable development can only be achieved with the support of the general public. In a post-industrial era with more and more environmentally friendly production technologies, it has been consumers' behaviour that has increasingly threatened the ecological balance. For sustainable development this means that not only the remnants of industrialisation have to be counteracted, but also the ongoing effects of consumerism.

It became important to educate people about the effects of their behaviour on the local environment, as well as to convey the interconnection of economic prosperity, social justice and ecological balance. A clear understanding of the interdependency of these three aspects had to be reached, in order to induce sustainable regional development. Here public environmental awareness is a prerequisite. Only when people begin to understand the consequences of their behaviour - that they threaten their own econom-

94

ic security by damaging the ecological balance - can they start to alter their attitudes and behaviour.

At the 1992 Earth Summit, the United Nations Conference on Environment and Development (UNCED) in Rio de Janeiro, 175 countries came to the same conclusion. According to the official UN definition, sustainable development is "development that meets the needs of the present, without compromising the ability of future generations to meet their own needs." In order to live up to this definition the representatives of the 175 countries signed the Agenda 21, a multifaceted plan for action towards sustainable global development.

Agenda 21 acknowledges education as a vital aid to support the changes needed for sustainability to take place. In a holistic view on environmental education, the awareness of the general public should be raised in order for a change of behaviour to take place.

Since then many local 'agenda-offices' have been working on regional levels to implement the goals that were agreed in the Agenda. However, as environmental concerns are not only of a global nature, but also affect local people's health and well-being, there are many other regional projects aimed at educating the public on environmental concerns.

It is important to convey local environmental awareness, as only people who know about nature will be able to protect it. Only a nature-aware society will be able to master future ecological challenges to ensure economically prosperous living standards.

The Role of Raising Public Awareness in Artery

Artery's aim was to regenerate post-industrial riversides. These riversides were often heavily neglected and sometimes heavily polluted. Ecological revaluation measures were taken, and previous public interaction with the river landscape was counteracted. Today the redeveloped sites are once again attractive spaces, accessible to the public.

Developments like the ten Artery pilots, whether they take place on public or private property, affect their surroundings in some way. It was therefore clear that the partners had to communicate with the public. This would improve acceptance among the population, and also ensure a certain degree of sustainability for the projects.

For the Projectteam Hollandsche IJssel it was important to educate local school children about the historic significance of windmills. Experiencing the rebuilt windmill Windlust and learning about the historic significance of windmills in the Netherlands raises the children's appreciation of their cultural heritage. They will be more sensitive to historic sites and their restoration, and will also come back to see the Windlust with their friends and family, contributing to its sustainability.

The Liverpool Sailing Club approached young people in the region to make them aware of the fascination of sailing. They aimed to bring adolescents back to the river and interest them in the nature surrounding the Mersey Estuary, where the Sailing Club is located. The young people came to understand the sensitive ecosystem of their region, as well as treasuring the river as a recreational space. At the same time this new regional awareness would also increase the number of Sailing Club members, who are active supporters of its operation and maintenance.

In some cases raising awareness led to active public participation in the planning and implementation of the Artery projects. In others raising public awareness was valuable for the later operation, maintenance and sustainability of the projects.

The river Neckar is one of the most important federal waterways in Germany, and therefore one of the region's most vital features. However, due to many alterations, such as river straightening and high bank enforcements, as well as infrastructure factors, such as main roads and railways, the Neckar is in many places almost inaccessible to the local population. Consequently the river has almost vanished from people's consciousness. Bringing the people back to their river and allowing them to experience a living riverside, is therefore a major goal of the development planners in the Rhine-Neckar region.

The project partner Nachbarschaftsverband Heidelberg-Mannheim, NV (Neighbourhood Association Heidelberg-Mannheim) had gained plenty of experience in raising public environmental awareness and education in the landscape development project

In Liverpool blocarting events raised public awareness for the new Liverpool Sailing Club facilities

'Lebendiger Neckar' (Living Neckar). This project was initiated in 1996. Since then investments in riverside regeneration were always closely connected with action days, events and projects along the river. More than 1,200 pupils from local primary schools took part in the specially designed teaching units 'Schulen für einen lebendigen Neckar' (Schools for a living Neckar). Due to this experience the Rhine-Neckar region was appointed thematic leadpartner for Artery's common strategy of raising public awareness.

With the pilot 'Agendapark Living Neckar' and the support of a river, the leadpartner aimed to support sustainable development through exciting educational programmes and various other awareness building events. The applied tools complemented the regeneration actions taken. The leadpartner has acknowledged that raising awareness was the key for public commitment. Through different types of educational actions and events - specially designed for both children and adults - implicit and explicit knowledge and experience were conveyed through emotional and practical learning.

A practical educational approach has to include bringing people back to their river to experience firsthand how important a healthy river landscape is for their quality of life. With the support of a 'river teacher' the Artery project partners have therefore developed different tools and methods in order to create sustainable public awareness for the region.

A Practical Approach
to Raising Public Awareness

To achieve long-lasting consciousness and commitment for a regional concern, a succsessful public awareness campaign needs to be thoroughly planned through. It should use a systematically structured approach to convey information and understanding to the general public, using a variety of communication tools. In order to choose the right methods and apply them correctly, several factors need to be taken into account.

Pupils enjoying the practical offers in the Neckar research laboratory on the museum ship in Mannheim

Establishing Goals
for a Public Awareness Campaign

Before the planning process can begin, the goals of the campaign have to be agreed. It is important to ensure that the goals are achievable and in some way measurable. If the goal has a very broad scope, it often makes sense to divide it into sub-goals. These often produce more visible results and ensure that motivation is sustained.

For the Rhine-Neckar region the overall goal is sustainable regional development. Such a goal needs everybody's support: public authorities, politicians, private companies, schools and other educational institutions, sports associations, environmental organisations and the general public.

The social and ecological importance of riversides is one aspect of sustainable regional development. The project partners Neighbourhood Association Heidelberg-Mannheim and the Verband Region Rhein-Neckar (Association Region Rhine-Neckar) therefore came to the conclusion that raising awareness for rivers and river landscapes had to be a sub-goal.

The Artery projects 'River Adventure Station' and 'Agendapark Living Neckar' with its integral project 'Menschen an den Fluss' (People to the River), as well as their forerunner project 'Schools for a living Neckar', are time-specific projects, and with their different actions to raise public environmental awareness have made a vital contribution to the overall target.

The individual activities within these projects had their own firmly formulated goals. The 'River Adventure Station', with its Neckar research laboratory, aimed to make children aware of the relationship between the information available onboard the interactive museum ship and the real river landscape outside. It put the children in a position to critically analyse the current developments taking place towards a sustainable riverside.

The goals of an awareness campaign express what is hoped to be accomplished in terms of awareness raising and public education. It is therefore important to consider how these goals can be evaluated later. Along with the campaign's aims, measurable indicators should also be developed.

In Stockport the UK Artery team aimed to create a sense of ownership for the newly formed Mersey Vale Nature Park. In order to achieve this goal various strategies to gain the support of the local public were applied during project implementation. Alongside these public participation strategies an educative awareness raising concept was employed. The goal was to create public responsibility for the new nature park. This would prevent the new facilities from being vandalised, as the locals (who included adolescents and children) would take pride in their environment.

Forming Key Messages

Corresponding to the goals of a public awareness campaign are the key messages to be conveyed. These messages can be explicitly formulated and stated, as in the value of natural habitats or implemented regeneration developments. In these cases the intention of the person communicating the message is clear.

But major implications can also be implicitly contained in other messages or actions. For example, the message 'regenerating river landscapes is worthwhile' can be concealed in the key message 'cycling along the river improves people's health'. The message 'investments in sustainable regional development projects contributes to the general quality of life of the whole local population' can be hidden in the message 'the new community park offers plenty of space for all kinds of activities for the whole family'.

Direct messages are easily accessible, but in abstract or more complex cases like, riverside regeneration and sustainable development, the indirect way of getting the message across is often more effective. This detour around an easy to understand message, can fertilise the 'hidden meaning' without the public being aware of it. The recognition then sets in through experiencing a change in perception. Keeping the possible consequences of a message in mind, it is important to carefully consider exactly what is being communicated to the public.

Determining the Target Audience

Once the goals are established and the messages clear, it is essential to consider who the awareness raising campaign needs to reach. Who is the target audience? Who else should be considered as a target group?

The selection of one or more target groups depends on the campaign's intentions. It is important to scan the population and determine which target groups are likely to be interested, and possibly already share the pursued perspective. Other considerations are community groups which, might be affected by the issues communicated in the campaign, and whether there are any special interest groups that should be approached?

When determining possible target groups, it is also strategically important to consider who might be opposed to the campaign's cause. In such cases it should be discussed whether these are potential audiences that ought to be approached, in order to affect their perception.

Understanding what each group values in terms of specific environmental or regional concerns, allows the message to be shaped and tools selected accordingly.

Selecting and Implementing
Awareness Raising Strategies

After determining the specific target groups, the next step is to select or develop one or more tools to disseminate the information to raise awareness.

For the success of the campaign it is vital that all information captures the audience's attention. As the social and demographic structure of a region is usually quite heterogenic, there cannot be a 'one-size-fits-all' solution. The selected tools should therefore respect group-specific needs and be in accordance with the range of skills, knowledge and interests of the public. For example, as a target group, children have to be approached differently to adults.

It is recommendable to create a work schedule for the individual activities, especially when employing more than one tool to raise public awareness. This plan outlines the different tools' single steps or phases, as well as individual responsibilities and a timeline. A work schedule simplifies the monitoring of the implementation process, and supports the later evaluation of the entire campaign.

Increasing public regional and environmental awareness requires a long-term campaign. This can be supported through individual actions and activities, but in order to create a lasting effect it needs continuous methods to communicate information and knowledge.

In co-operation with a 'river teacher', the Artery partners in the Rhine-Neckar region pursued a multifunctional approach to activate a broad public with different levels of knowledge and diverse interests. This variety of tools also helped to ensure a continued rise in public awareness for sustainable regional development. In the project 'Menschen an den Fluss' (People to the River) different 'modules' were employed offering people the opportunity to rediscover various facets of their river. Through different events and educational programmes relating to nature, history, economy and culture, people were brought back to the riverside.

Precondition: External Support

When planning a broader campaign, which will take place over a longer period of time and consists of various tools, it is beneficial to look for support within the region. Partners in the public as well as private sectors are often able to support individual activities with expertise, service and equipment. Drawing the partners together in a topic related network helps to ensure the continuity of the awareness campaign.

In order to fully implement the different modules and secure their continuation, the Artery partners from the Rhine-Neckar region initiated the regional 'Neckar-Netzwerk' (Neckar Network). This network consisted of environmental organisations, specialised authorities, institutions, local authorities and professional individuals.

The 'Neckar activists' enhanced the workflow of the individual modules on-site. They gave input to the educational programme and different events. Thanks to the partners' diverse backgrounds, it was possible to design and realise a multitude of educational services for different target groups. The established educational programme, named after the project 'People to the River', ranged from specialist nature tours and art classes along the river, to activities conveying information and experience on river ecology.

In addition to 'People to the River' in the education centres, the project team was able to offer specialised tours on and along the river Neckar. This was only possible due to the support of the Neckar Network. In co-operation with the local canoe clubs, the river expert offered guided canoe tours on the Neckar. The

view from the water of the river and its banks gave participants a completely new perspective on one of the regions' main features. In return the network participants were able to promote their own work to the public at the different events.

Environmental Education

The overall target of environmental education is to promote global awareness, sustainable living and active public participation. The assumption behind environmental education is that once a person is properly educated and sensitised regarding the environment and sustainability, his abilities and willingness to act will develop accordingly. Environmental education is therefore an important approach to increase public awareness.

Since the signing of Agenda 21 the member countries of the European Union have recognised environmental education as a significant method to increase environmental awareness, encourage public participation and provide the basis for rational decision making on environmental issues. At the international con-

ference for 'Environment and Society: Education and Public Awareness for Sustainability' at Thessaloniki in 1997, the definition was broadened to 'education for environment and sustainability', as it is possible to transfer the goals of the Agenda from general environmental concerns to those of sustainability.

Education for environment and sustainability has therefore become an important element in the schools' curriculum. Children as future decision makers have to know how regional ecosystems work and be sensitised to environmental concerns. There are also many public education programmes aimed at adult target groups, because their decisions and behaviour effect the current environment, not only tomorrows.

In order to reach both adults and children, the Neighbourhood Association Heidelberg-Mannheim and the Association Region Rhein-Neckar initiated the implementation of educational programmes for children as an investment in the future, but also with the intention of reaching the children's parents and rela-

Partners of the Neckar Network on a fieldtrip along the Neckar

tives. A diverse programme for adult education was also designed.

Having acknowledged its significance, environmental and regional education played a major role in Artery. Recognising environmental education as a cross curriculum topic, it was their goal to develop interdisciplinary environmental education programmes for different levels (primary schools to adult education), as well as extracurricular courses and events. The formal school curriculum would be complemented by the emotional and practical approach that was adopted in many of the programmes on offer.

Schools as Multipliers for Awareness

Schools are important multipliers in the process of creating awareness and gaining knowledge. It is advisable to place the emphasis on environmental education in an awareness raising campaign for schools and teachers.

In the Rhine-Neckar region the tools designed for school children and teachers, were based on the successful completion of an earlier project, 'Schools for a living Neckar'. These tools were adapted and extended in the course of the Artery project 'People to the River'.

School Event Days

One way of increasing the children's environmental knowledge is through event and action days organised for school classes. Offering a collection of educational events to choose from over several days, is a good way to get in contact with the schools in the region and trigger specific demand. These days should be led by an expert in the field and take place on-site, close to the schools, to ensure that the children realise their 'outdoor classroom' is part of their everyday environment.

The different activities in the 'Schulen an den Fluss' (Schools to the River) action days, in the Mannheim

Children and adults analysing Neckar water for aquatic animals and plants on the 'Action Day Living Neckar' // Children building works of art following the example of 'Land-Art' during the event days 'Schools to the River' in the Mannheim and Heidelberg area

Consider Level of Experience

In order to design an attractive educational programme, it is important to first determine the existing level of experience and awareness. Otherwise you are in danger of being boring, or expecting too much from the target group. This can easily be the case with children, who might then lose interest in the whole topic.

and Heidelberg area, were attended by experts with specialist knowledge. During the Artery programme the event days took place three times. For five days local primary school classes were able to participate in up to 50 exciting events on different topics relating to the river Neckar. In the 'River Expedition' along the Neckar, the children were able to discover the river habitat with all their senses. In a water expedition, micro-organisms living in the river were tracked and the children learned about their vital role in the local ecosystem. With material taken from the river, the children built works of art following the example of Land-Art. All events gave the children time and space to make their own discoveries about the environment on their doorstep. The events conveyed the ecological importance of the river and its banks, and the recreational joy to be gained in spending time in local nature.

Developing Special Teaching Units

Another opportunity to involve schools in a campaign for raising public regional and environmental awareness, is by developing corresponding teaching units. In most European countries environmental and regional education is integrated in some way, into the different curriculum subjects. This can often be enhanced from 'outside'.

The Artery partners in the Rhine-Neckar region knew from experience that, when it comes to new subject matter, most teachers welcome support from nature specialists, especially regarding outdoor activities.

In the course of the 2004 educational reform, the Federal State of Baden-Wurttemberg initiated a new interdisciplinary subject 'Mensch, Natur und Kultur' (Man, Nature and Culture), or MeNuK. This new subject aims to encourage the fascination of scientific phenomena as well as technical connections, and to broaden the children's creative approach to arts and music. MeNuK's overall goal is the systematic introduction and mediation of the region's cultural heritage. This interdisciplinary approach offers teachers the chance for practical and problem-oriented, active and creative lessons. Certain parts of these lessons are held outdoors.

Although material for the new subject was available, teachers were rather reluctant to take their classes outside to the Neckar. In co-operation with many partners from the Neckar Network, the Artery project team aimed to support local primary schools in the implementation of the new interdisciplinary subject.

The 'river teacher' working for the Fortbildungsgesellschaft für Gewässerentwicklung mbH WBW (Training Association of the Baden-Wurttenberg Water Management Association), the local institute of further education for inland water development, carried out several special teaching units with primary school classes in Neckarhausen. The WBW works on behalf of the Baden-Wurttemberg Ministry of the Environment and was founded by the Baden-Wurttemberg water management association. The institute is a vital partner and initiator of the Neckar Network.

The special teaching units for children from grade 2 to 4 aimed to bring the Neckar as a natural habitat closer to the children, and served as instruction for the teachers. After this training the teachers felt confident taking the children out to the local riversides and using

them as 'natural classrooms'. As a further teacher's aid the new teaching units were designed on the basis of MeNuK's curriculum. Different regional and environmental topics were incorporated to ensure practical-oriented and exciting lessons.

The training carried out by experts, as well as the supporting material, was very much appreciated by the teachers. Training the teachers is an important tool in raising regional and environmental awareness. This combination of factual information and educational instruction enhances the teachers' awareness, as well as provides them with a method of conveying knowledge and experience to their pupils.

Awareness Raising in other Educational and Cultural Institutions

Although schools are a main multiplier for raising awareness, similar educational programmes can also be made available to children and young people outside the curricular context. Such services could also include a general adult target group.

Public Education Programmes

Public education centres offer a vast range of recreational and further education classes for children, adolescents and adults. People use these facilities in their free time for personal development as well as recreation.

Designing classes for public education centres and other educational institutions offers the opportunity to include diverse courses and events to create regional and environmental awareness. This diversity attracts different target groups and ensures that a broad public is reached.

In co-operation with the Neckar Network, the Artery partners realised a broad spectrum of courses for both young and old. 'People to the River' was carried out with the public education centre in Heidelberg and the Abendakademie Mannheim (public education institution). This programme included guided tours relating to the history of the river Neckar, events regarding high water protection, and sculpture classes using flotsam from the banks of the river.

Adolescents and adults in Heidelberg gained new sights of the river landscape along the Neckar on a 'mirror tour' within the context of the public education programme 'People to the River'

Museums

Other institutions where environmental and regional education takes place are museums. It is therefore advisable to include local museums in the awareness raising campaign.

In co-operation with the Landesmuseum für Technik und Arbeit in Mannheim (State Museum for Labour and Technology) the Artery partners aimed to sensitise the public to the Neckar, on-site.

The 'River Adventure Station' and its exhibition 'Was(s)erleben – Entdecke den Neckar' (Water experience/water life – Discover the Neckar) offers children an environment where they can find out all about the river as a habitat and waterway. Due to the position of the museum ship, children are able to transfer their newly found knowledge from the hands-on exhibition, to the real nature outdoors. Changing events invite children to visit the 'River Adventure Station' again and again.

Accompanying Riverside
Regeneration Projects with Educational Activities

Schools and other educational institutions are important multipliers for raising public awareness in children and adults. Nevertheless, it is also important to promote regional and environmental awareness, on-site. Regional development projects provide excellent opportunities to enhance the public's awareness of regional concerns.

Tools for raising awareness for individual developments need to focus on the specifics of the implemented measures. Activities are dependent on the nature of the project and are often single short-term events to make the public familiar with the enhanced sites. In some cases continuous measures are also appropriate.

The benefit of raising public awareness for specific sites is that the effects go far beyond the individual projects. Regional development projects strengthen regional identity, which in turn increases the public's identification with their region. When the public's curiosity has been raised for a certain regional issue, it can be assumed that people's awareness for related topics is also increased.

Tools for raising awareness for special sites are usually publicity-effective activities or events, like exhibitions, guided tours or rallies for children. Educational features can also be implemented at the regenerated sites.

Educational Programmes for Individual Sites

In order to enhance awareness for regional cultural heritage, the Dutch Artery partners developed a special educational programme about the restored Windlust. Brochures were developed for a target group made up of pupils from four highschools in Nieuwerkerk aan den IJssel. In order to make the material easily applicable, a separate teachers' brochure was also developed. Since the Windlust's complete restoration, the teaching units can now take place in the windmill facilities.

The Windlust educational programme also contains information about other windmills and their general historical significance. The benefits resulting from educating children on-site have been increased support from the local public for the windmill's restoration, as well as a growing awareness of the importance of this cultural heritage.

Raising children's awareness for individual sites has the additional effect that once their interest has been raised they will later return with friends and family. This increases the range of people reached by on-site awareness raising tools. In the case of the Windlust, it is possible to visit the mill several days a week. On these days the 'Molen Kortenoord' foundation, the project initiators and Artery project partners, offer guided tours through the mill. An exhibition on historic mills and flyers, enhance the visitors experience.

Promoting understanding about the region and consequently inducing a sense of ownership for public spaces, can also be achieved by seeking the public's opinion on certain issues.

For the Mersey Vale Nature Park, the project partners MBC and Stockport Council commissioned artwork for the park's entrance. In order to raise local children's awareness of the newly built nature park, the artist carried out several day-long workshops with a group of 9–10 year old pupils from a local school.

Through a telescope children can compare the exhibits on the museum ship with the real riverside just outside the window // Pupils caught fish to learn about their anatomy and categorise them accordingly on prepared charts during the project 'Schools to the River'

During the workshops they visited the nature park to gather ideas and materials, develop their ideas using different styles of artwork, and produced large-scale drawings to help them visualise how their artwork would look in real life. They also listened to stories from a few local residents, who had lived in the area all their lives and were able to give some historical background to the area. By engaging the school children in this way, they became more aware of their surroundings and the nature park. They also felt a great sense of ownership for the installation that the artist created from their drawings.

Encouraging Public Dialogue through Competitions

Tools that encourage people to critically analyse their surroundings are an important instrument in raising regional awareness. The Artery partners in the Stuttgart region therefore designed a photo competition. Through the media people were encouraged to identify and take pictures of spaces by the Neckar that they found especially attractive or particularly unattractive and in need of regeneration. The best photos won an award.

Artery's public awareness leadpartner accompanied the regeneration projects with various attractive educational events. The shallow water zone in Ilvesheim, near Heidelberg, was opened with a nature expedition along the river. Likewise in Schlierbach the mayor officially opened the site with a high-profile event attended by local people and schools.

Activities like this brought the newly developed sites to the attention of the public and increased their identification with the region. The measures taken to create awareness for individual sites had the additional benefit that people came to understand the need for regeneration investments. Local authorities were thereby able to justify the often cost-intensive projects.

Children of a Stockport primary school designed the layout for the artwork for the entrance of the Mersey Vale Nature Park. Through this involvement their awareness and sense of ownership for the park is long-lastingly increased

Promoting Public Awareness through Publicity

Another important tool in raising public awareness is creating publicity for certain issues or developments. The more people become familiar with these concerns, the more they are affected by the campaign. Publicity is an important tool in creating awareness and consequently triggering a change in behaviour. It also ensures the success of other awareness raising methods. If an educational institution programme is not made public, it will not be reach the people. Therefore, different publicity gaining tools are necessary for any campaign seeking to raise public awareness.

Public Relations

Newspapers, radio and television reach thousands of people in a region. Therefore, they play a significant role in a campaign. High media coverage guarantees that the key messages are communicated to a broad public.

The Artery partners made a great effort to ensure regular media coverage for the ten pilots, which includes the offers of the 'river teacher'. In the five regions press releases were sent out to the regional media to inform the public about the progress of the project implementation. Additionally, local journalists were supplied with background reports about the projects and their regional significance, and were regularly invited to background talks with different project partners. Good contacts with the local press can ensure maximum media coverage.

To ensure that people knew about the educational programme 'People to the River' and the awareness raising events related to the local Artery pilots, the project partners in the Rhine-Neckar region supplied the regional media very early on with relevant information on content and starting dates. Further flyers and brochures were distributed at relevant places to ensure maximum awareness of these activities.

Internet

As well as establishing a good relationship with the press, being present on the internet can also gain publicity. In Artery the 16 project partners made sure that their pilots were promoted on their websites. Artery additionally maintained a website promoting the ten pilot projects as well as their transnational co-operation. The website has proved to be an important communicator of publicity on a local, as well as the international European level.

Events, Exhibitions and Fairs

Events are a popular tool to raise awareness. They guarantee high media coverage and attract many people. There are two ways of using events for publicity purposes: either use existing events, or create your own.

Using existing events has the benefit that they are already established and well-known to the public. On the other hand, it is more difficult to gain people's attention on days where many organisations are presenting themselves and are offering different kinds of activities. When creating your own events, enough lead time has to be allowed for.

In Capelle aan den IJssel the Artery partners initiated a public barbeque at the site of the former shipyard, to promote the regeneration of the land into a community park. The Projectteam Hollandsche IJssel thereby raised awareness and facilitated the active participation of the local public. Three years later, after finishing the redevelopment, the new Vuykyard park was officially opened with a grand event. This event increased the local population's awareness of the new park and its relevance to their quality of life.

The highlight of the opening was a collective 'bench-sitting'. People were encouraged to sit on a knee-high wall along the boulevard. More than 300 people found a place on the longest 'bench' in the world. For this event Capelle aan den IJssel gained an entry in the Guinness Book of Records. Every 'bench-sitter' received of photo of the event, which symbolically shows the outcome of a public participation initiative. This event created a lasting impression on the people, and made them value the park and care about its maintenance.

The project partners of the Artery pilot 'Barrier-free Tourist Attractions' in the Ruhr Valley used a very different approach to increase awareness for the newly implemented RuhrtalFerry. They offered locals the chance to take the ferry operator's driving test, and take on occasional shifts as voluntary ferrymen. This attraction increased publicity for the RuhrtalFerry and contributes to its sustainability, because more people are now using the ferry and enough honorary ferrymen guarantee its operation.

Success Story: Regional Awareness ensures Project Implementation
One of many events relating to the individual pilot projects was a transnational event called 'Wasserzeichen' (water signs), hosted by the Artery team in the Rhine-Neckar region. It was the final event of the 'Menschen an den Fluss' (People to the River) project. For this event, children from a local primary school prepared artworks for a river exhibition.

On a solar catamaran the European Artery delegation, public and private partners of the Rhine-Neckar projects and local politicians, as well as the participating school children, took a tour of the regenerated riversides from Heidelberg to Mannheim, that were part of the Artery pilot 'Agenda Park Living Neckar'.

The publicity gained through this event, its supporting programme and the children's art exhibition, promised to further increase regional awareness. Therefore, the project partners ensured that in the run-up to the event huge publicity was raised for the event, and for the individual local Artery projects. In the end this ensured that all partners and authorities kept to their schedules, and all projects were realised in full. Public awareness can therefore also have a controlling function.

Evaluating the Results
of Raising Public Awareness

The last step of a public awareness campaign is to evaluate the results. This is important for two reasons: it needs to be determined whether the established goals have been achieved, and it is important to evaluate the usefulness of the applied tools.

Awareness is a psychological factor and is not easy to measure. In the case of raising public awareness for a specific project, evaluation can be gained by monitoring visitor numbers. Another indicator is the way people treat the sites. A campaign has been successful if places that were previously vandalised are being properly utilised post-regeneration. People are treasuring it and taking care that it is well maintained.

In the case of raising awareness for sustainable development, the evaluation of actions is more difficult. How can you measure if people's attitude towards local nature has improved? How can you determine if educational services have increased people's knowledge of the river as a natural habitat and social source of recreation? To a certain extent these aspects can be evaluated in a similar way to awareness for individual developments.

One indicator of strengthened public awareness is an increase in public participation. The tools of a campaign for public regional and environmental awareness have been educatively successful, if they have increased the public's respect and responsibility for the key features of their region, be it cultural heritage sites or landscapes. The increased sense of responsibility then leads to an increase in active commitment.

An important indicator of whether individual methods to raise awareness were successful can be measured by monitoring how well these tools were accepted by the public. In the case of events, exhibitions or educational projects, visitor numbers are also decisive. If programmes are not frequently visited, although they have been made public through the media, they are unlikely to have influenced people's attitudes. While frequently visited programmes have a good chance of reaching the public.

With the educational programmes 'People to the River', the action days 'Schools to the River' and 'Man, Natur and Cultur on the Neckar', as well as the educational measures taken to accompany individual regen-

Raising
Public Awareness

~ In order to get started it is important to first establish goals. Remember that in order to evaluate the campaign later, you will need indicators to measure the success.

~ Identify target audience: Determine who is important to be reached.

~ Formulate Key Messages: Shape the messages that you aim to communicate to the public according to the specific target groups.

~ Select tools that suit your target group and key messages.

~ In order to implement the selected actions, find external support and keep an eye on the campaign working schedule.

~ Evaluate results to ensure that your campaign has been successful. Take time to consider, which tools proved to be more efficient than others.

eration projects, the leadpartner for Artery's strategy public awareness brought more than 2,500 people to the Neckar. By May 2006 the museum ship had seen over 8,970 visitors. There is a good chance that many of these people now see the main feature of their region, the Neckar, through different eyes.

With a great opening event and an entry in the Guinness Book of Records, a lot of publicity was drawn to the new Vuykyard community park

Text: Sarah Wallbank

Public-
Private
Partnership

The windmill Windlust and the Speke and Garston Coastal Reserve: Two of the public-private partnership highlights of Artery

Introduction

In the last two decades the phenomenon of public-private partnership has made an entrance into local and national public authorities across Europe. Public-private partnerships are strategic partnerships between public and private institutions, working together for the implementation of public projects.

This identification of the private partner with a public project generates broad support, and furthers its realisation with financial and in-kind contributions. In such a joint effort, it is also possible to establish a public-private partnership to operate or maintain a public facility. The private and public partners share the risk and responsibility for the successful implementation and the sustainability of the undertaking. However, working together in a partnership, the public and the private sector are often able to achieve more than the sum of their individual capacities. So far the fields most common for public-private partnerships are infrastructure, urban development, health care, schools and supply and disposal services.

These numerous examples of the successful involvement of the private sector in public domains inspired the Artery partners to include the strategic approach public-private partnership in their common themes and strategies.

For the five European partner regions, public-private partnerships have proved to be a useful and beneficial instrument. However, in order for such a collaboration to be successful many factors have to be taken into account.

What is Public-Private Partnership?

Public-private partnerships are a form of collaboration between national, regional or local public bodies and private companies to meet the needs of the public. The involvement of private businesses in the public sector can have many forms and encompasses anything from sponsoring to the management of whole services.

Benefits of Public-Private Partnerships for the Public Partner

~ Private partners bring in their own expertise, which facilitates the effectiveness of the workflow and can accelerate the delivery of the project.

~ Resources can be saved and used for other investments in the project, for example in-kind contributions, monetary or human resources.

~ Through joint efforts results are produced that are beyond the capabilities of the single partners.

In the Artery programme public-private partnerships have contributed to the realisation of most of the ten pilot projects in various forms and to different degrees. The accomplished partnerships comprise in-kind contributions, like the special rates for the bikes needed for the barrier-free tourist attractions in the Ruhr Valley, which were granted by a local bike retailer. And monetary investments, such as the private sponsorship of particular elements for the windmill 'Windlust'. In some cases the sustainability of the projects was guaranteed through management and maintenance contracts, or trade and service agreements. For the Speke and Garston Coastal Reserve the project partners, Mersey Basin Campaign (MBC) and Peel Holdings, co-founded a special management company. This public-private company shares the responsibility of the ongoing management of the reserve and helps to attract new partners. In the Ruhr Valley several private partners founded the operating society ‚Unser Freibad am See e.V.' (Our Open-Air Swimming Pool at the Lake Association) which now takes care of the River Bath Ruhr.

In a public-private partnership the partners agree to work together to achieve common objectives in a well-defined timeframe. A formal contract or an informal agreement (letter of intent), usually specifies the partners' individual responsibilities within the project. Artery however, demonstrates the flexibility of the public-private partnership approach. In several of the pilot projects public-private partnerships were established without any form of contract. All parties involved strongly believed in the project and put its success first. It was therefore possible to work in partnerships without a legal framework.

Besides securing extra-funding for public services, another benefit is that in the long run public-private partnerships can enhance efficiency. The results then often exceed the initial goals. However, one must also bear in mind that establishing trusting partnerships takes time and effort on both sides. It is also important to make sure that all partners gain from the partnership. Creating a win-win situation is a solid foundation for success.

The Role of Public-Private Partnerships in Artery

In its twenty years of existence the Mersey Basin Campaign (MBC), Artery's UK thematic leadpartner, has gained plenty of experience in working with the private sector. On many occasions different forms of public-private partnership have enhanced the implementation of numerous projects.

Through this past experience the Artery partners were convinced that such collaborations could facilitate the pilot projects. They therefore decided to make use of the public-private partnership approach in their own way.

The regional Artery partners hoped to establish local and regional partnerships for riverside development in the five European partner regions. The partners saw the most efficient means of gaining support for the successful full-scale implementation of the different projects, through a public-private partnership approach. The lead- and project partners compiled a wide scope of co-operation from public, public-private and public-NGO partnerships.

Benefits of Public-Private Partnerships for the Private Partner

~ New business opportunities

~ Access to the public sector

~ Good working relationships

~ Positive profile and publicity

Thanks to public-private co-operations the Liverpool Sailing Club finds a home in a new state-of-the-art building // The water playground in Heidelberg under construction. It could only be finished with the support of a private partner

Public-Private Partnership

Strong and lasting private sector involvement can also offer sustainable operation and maintenance of the new facilities. Artery aimed to make the regional public authorities aware of the potential gained through public-private partnerships, as well as making local public and private bodies aware of the benefits of working together for riverside regeneration.

Public-private partnerships provide opportunities for integrated thinking through combining varied perspectives from all sectors, which would help Artery achieve their main goal of creating new economic and social opportunities in an ecologically sound environment. The five regions therefore decided to promote local and regional partnerships to involve stakeholders, private enterprises and decision makers in the projects during the early stages of the planning process.

The strategy of public-private partnership presents great economic potential for the Artery pilot regions. It facilitates the regeneration of urban riversides, while increasing business opportunities for local companies in the initial phase of project implementation. Public-private partnerships also create long-term opportunities for the local economy, due to operating and maintenance requirements, as well as the positive economic consequences of the projects.

Therefore, Artery's key message for public-private partnerships was that businesses and riversides can thrive by working together. An investment in Europe's rivers is an investment in an economically prosperous and ecologically sound future.

Artery as a Learning Partnership

Although increasing numbers of authorities recognise the potential of public-private partnerships, their use has not spread to all European regions to the same extent. It is therefore not surprising that at the beginning of the Artery programme the five partner regions held different levels of expertise.

When Artery started, the partner regions' knowledge of public-private partnerships clearly reflected its use in the respective countries. The UK has the most experience in public-private partnerships, because this is where the concept originates. Other European countries, like the Netherlands and Ireland, soon picked up the idea and started their own experiments in collaborating with the private sector. At the beginning of the 21st Century

Germany had not gathered a lot of experience with public-private partnerships. Since then federal public-private partnership task-forces have been established and initial projects have been successfully realised.

Today in Great Britain over 20 percent of all public investment projects are realised with the support of the private sector. One of the earliest public-private partnership projects still in operation is the Mersey Basin Campaign, the UK Artery leadpartner itself. The Campaign was established in 1985 as a regeneration programme for England's Northwest. It was a direct consequence of the riots in Liverpool, which started four years earlier. In a joint venture the government, private and voluntary sectors set out to address the problems of water quality and landward dereliction of the river Mersey and its tributaries. It was one of the first public-private partnership approaches, which combined economic prosperity and the quality of the environment. This interdependence is reflected in the Campaign's three key objectives.

Since its founding, the Campaign has developed from a government-run initiative, led by an independent chair, to its current partnership status. Its industrial members, like Shell and Unilever, are encouraged

The three Mersey Basin Campaign key objectives are

~ Improve river quality so that all rivers, streams and waters in the Mersey and Ribble catchments are clean enough to support fish by 2010.

~ Encourage riverside regeneration.

~ Actively engage the public, private community and voluntary sectors in the process.

to incorporate the Campaign's objectives into their daily activities and business practices. In this partnership model Mersey Basin Campaign brings disparate groups and sectors together to work towards a shared mission.

Having worked with numerous private partners of different sizes, the Mersey Basin Campaign knows what it takes to establish successful partnerships. Therefore, the other Artery partners chose the UK organisation as the thematic leadpartner for the common public-private partnership strategy. Through meetings and workshops the Mersey Basin Campaign shared its private sector knowledge with the Artery partners.

Practical Approaches
to Public-Private Partnership

In the last few years various public authorities, such as regional and national ministries and task-forces, have developed an understanding of public-private partnerships. They derived their experience mainly from larger projects in infrastructure related areas. However, these proposed definitions and strategies can also be successfully adapted to smaller undertakings, as well as new areas of public-private partnerships. The partnerships that were established in the Artery pilot projects are therefore ideal to demonstrate what it takes to create successful working relationships.

Identifying Possible Partners
for Public-Private Partnerships

Prior to involving public or private partners in the implementation of a project, it is important to clearly define the desired goals. The planning phase is essential for building public-private partnerships. It is often the earliest moment to think about how the private sector, or other public bodies, can become involved in the project. Essential contributions needed from the private sector are often financial support and entrepreneurial input. For some projects, in-kind contributions are necessary. Others require a partner's continuous expertise in the construction or operation of a project. As every project is different it is therefore important to put down the specific requirements of the respective endeavours.

Once the suitable form of public-private partnership has been determined, or which forms could facilitate it, the next step is to find out whom to approach.

Public-Private
Partnership Essentials

Committed partners

A public party:
~ municipalities
~ legislative bodies (courts, politicians)
~ executive bodies (governments, administrations, authorities)
~ any other public institutions

A private party:
~ private companies
~ economic associations
~ labour unions
~ chambers of commerce
~ interest groups
~ private/single individuals
~ clubs and societies, citizen action groups

At least one player from each side is needed to form a public-private partnership!

It does not make sense to start without a plan and contact just any business in the region. This kind of approach would be unsuccessful. In their strategy the Artery team agreed on four aspects that should be accounted for when considering whom to approach.

In order to find the appropriate companies the project initiators must scan their economic surroundings, find out who has the suitable resources and a vested interest in seeing the project successfully realised. Companies that see an economic benefit from the implemented project are more likely to be interested in investing than others. The same is true for companies that share common interests with the project developers. They often benefit from the realised project, for example from the development of land. In a partner search these organisations should be identified and contacted.

Four Aspects
to Consider
For Whom to Approach

~ Who has to be contacted? (i.e. landowner)
~ Who suits the projects? (i.e. construction companies or service businesses)
~ Which companies are situated close to the project?
~ Who would benefit from involvement in the project? (i.e. local businesses, clubs, private interest groups …)

With these considerations in mind the Artery partners in the Rhine-Neckar region searched for private partners who would be interested in investing in the planned water playground on the banks of the river Neckar in Heidelberg.

Being Artery's thematic leadpartner in common strategy awareness, the implementation of the water playground was an integral part of the pilot project Agendapark Living Neckar. Playing with water helps children to understand the laws of physics and brings them closer to the river. The playground facilities were designed to resemble the flowing river and were to be given a natural style layout. The water for the facilities would be gained from a groundwater fed well. So when the real costs of the construction of the playground turned out to be much higher than anticipated the whole undertaking was endangered. It therefore became essential to find private financial support.

After considering who might be interested in helping with the realisation of the water playground, the Rhine-Neckar project partners decided to approach organisations with an interest in water. They found a suitable partner in the Stadtwerke Heidelberg AG (Public Services Heidelberg Corporation). The local water supplier had a clear interest in the playground, because the new facilities would help to raise people's awareness of the community's water supply. To support this Artery project the company contributed the needed 100,000 euros for the construction of the playground.

The situation was similar in the second Artery region in Southern Germany. The project partners in the Greater Stuttgart region were looking for private investors with an interest in riverside regeneration. As the pilot Wernau Neckar-River Reserve had predominantly ecological goals, they searched for private partners who shared these interests.

The sluice in the river Neckar near the community of Altbach presents an insurmountable obstacle for the upstream migrating fish. Creating a tributary around the sluice would provide an ecologically sound alternative for the sensitive fish population. Within the context of Artery, a feasibility study examined if such a fish pass could be established by reconnecting an old tributary with the river. A newly built creek could connect the old tributary with the cooling canal of the power plant operated by the EnBW Kraftwerke AG and thereby create a stream around the local hydroelectric power plant. For such a project the Artery project partners needed private partners to provide expertise and land for the implementation of the fish pass.

Power generating companies are renowned for using a lot of natural resources to produce energy, and to the residents they often present a disturbing view on the landscape. It is therefore particularly important for energy suppliers to take actions towards ecological regeneration. Local power stations in particular aim to create a sustainable positive image among the population, to state their commitment to the region and prevent clients from turning to different providers. The Artery partners therefore presumed that EnBW might be interested in contributing to a 'green' project in their local vicinity. Being part of such a partnership would give the company's image a long-lasting boost within the region's population. Consequently the EnBW was contacted to propose a public-private partnership.

The energy provider contributed to the feasibility study by offering the services of two of the company's experts for environmental concerns. They supported the project with their advice regarding the power plant operation and the implications that the proposed fish pass might have. In the event of a positive outcome,

Choosing the Right Partner

~ Consider what the partner can bring to the table.

~ Ensure the partners have the capacity to deliver their responsibilities. Shortages can delay or endanger the whole project!

~ Check the partner's true motivation for supporting the project.

~ Consider carefully the partner's policies and objectives and make sure that they mostly correspond with your own. If objectives clash, the whole project could fail (i.e. a lack of public credibility can lead to loss of acceptance of the project. The withdrawal of an essential partner from the partnership could result in the whole project team being unable to deliver.).

~ Remember: choosing the wrong partner makes the whole partnership and project implementation more difficult. Therefore, intensive discussions and negotiations are essential to avoid a negative surprise!

Water playground Heidelberg: the residents profit from the public-private co-operation

the energy supply company EnBW would also make its land available for the realisation of the fish pass.

An Early Approach to the Private Sector

Extra funding or other support from the private sector (i.e. human resources) secured early on, can be included in the project scheme and the allocation of means distributed accordingly. Therefore, if possible, private partners should be included in the planning phase of a project.

In the pilot project Speke and Garston Coastal Reserve at the Mersey Estuary in Liverpool, the Mersey Basin Campaign (MBC) approached the private concern Peel Holdings at the very beginning of the project scoping. This was necessary as Peel is the respective landowner of the reserve and without the company's consent, there would not have been a project on this location. However, Peel also provided means for legal advice and management resources throughout the implementation of the project. Additionally Peel paid for the round-about at the entrance of the Coastal Reserve, as well as for the artwork, which was acquired to enhance the aesthetic interplay of nature and culture within the vicinity.

In return, the property developer profited from the enhanced land as two of his important assets bordered onto the Coastal Reserve; the Liverpool John Lennon Airport and the newly constructed business park. With a direct view of the heavily deteriorated land, it was difficult to convince new businesses to locate there and lease land. The beautiful Coastal Reserve promotes the site and helps to attract businesses. Ultimately, higher land quality allows for higher rental prices. Peel's contribution to a public green space on their private land also increases their image among the population. The beneficial effects for both MBC and Peel Holdings were maximised, because the partnership was established right at the beginning of the project planning phase.

The work of the Artery partners shows that it is not always the results of a project that can inspire businesses to become involved in a public project. Sometimes

the planning phase is more essential for the company than the implementation or operation of a facility. The prospect of being able to negotiate favourable conditions in return for their contribution makes companies seek an active role in the planning process.

With the planning of the Mersey Vale Nature Park in Stockport, Emery Farm Holdings profited from forming a close working relationship with Stockport Metropolitan Borough Council. The company is the owner of the former bleach works site that has lain idle for many years. Artery partners including the MBC in Stockport and the Stockport Metropolitan Borough Council favoured the idea of integrating the land into the Mersey Vale Nature Park.

In negotiations Emery Farm Holdings was able to participate in the planning of the new nature and community park. In the established partnership it was then possible to negotiate mutually favourable planning conditions for both sides. Emery Farm Holdings agreed to donate the land forming a part of the nature park. In return, the company was permitted to distribute soil on the site, thus resolving a potentially expensive operation.

This public-private collaboration would not have been established if all planning issues had already been solved. In order to win the support of private partners, it is therefore important to clearly point out the opportunities for companies and organisations that arise from taking an active role in the planning phase.

Approaching the Private Sector

Once the potential partners have been identified they must be contacted to promote the project.

Contact through Established Partnerships

The most convenient way of approaching a potential partner is through already established partnerships that have been built through other projects. The great advantage is that both parties already know who they are dealing with: contact people, priorities and business objectives, as well as partners' working methods are already known and a level of confidence has already been established. This alleviates the process of making new contacts and is a good foundation for motivating the private company to become involved in the project in hand.

Private land of the local power supplier in Altbach; on this ground an ecological regeneration project will be realized // Derelict land of the Coastal Reserve at the Liverpool Sailing Club. Before the local public-private project, this area was an obstacle for the adjacent business park

In Artery, the project partners of the pilot Mersey Vale Nature Park were able to use such an existing partnership. The Mersey Basin Campaign (MBC) has been working with the service provider United Utilities for years. For the Artery pilot the contact was made through an MBC Policy Advisor working on secondment from United Utilities.

For the Mersey Vale Nature Park several sites were to be drawn together. Parts of the park were former sludge beds belonging to United Utilities. Through the established relationship United Utilities were convinced to keep these sludge beds out of operation and pass their management over to MBC. If the company needed the land in future, the MBC would offer them neighbouring land in exchange. The service provider probably would not have agreed to the deal if there had not already been an existing, trusting relationship.

Contact through Community Leaders

When attracting potential partners, it is important to give credibility to the project. This is especially necessary when approaching larger companies. Increasing ones own credibility can often be achieved if the contact to the private partner is made by a well-known public representative or patron. Besides adding to the project's credibility public representatives can also help to arrange contact with the right people.

In order to gain the support of the energy supply company EnBW in Altbach, the Artery team in Stuttgart won over the local mayor, who used his position to promote the significance of the fish pass. The mayor is an important public representative in the community, and the municipality is an important client of the energy provider. His support of the ecologically and socially valuable project helped to convince EnBW's decision makers to back the proposed undertaking. Since the initial contact through the mayor, the large energy company has been actively engaged in conducting the feasibility study for the fish pass on the grounds of Altbach's power plant.

Former bleach work sites at Stockport: Thanks to early involvement of the landowner, they became part of the Mersey Vale Nature Park

Contact through Unconventional Means

In the competition for the private sector's attention, it has also proved beneficial to come up with unconventional ways to initiate contact. This is particularly essential if there are no existing connections to potential partners.

This was very much the case for the Artery partners in the Ruhr Valley. In order to get in touch with local businesses, the regional Artery partners Witten Labour and Employment Company (WABE) and the Ennepe-Ruhr County hosted a grand gala dinner for key local businesses in the region. From this gala dinner emerged the 'Ruhrtafel' Circle, which hosted periodical events where business representatives and regional political decision makers meet.

For the Ruhr Valley this was a brand new way of approaching potential sponsors and partners. During the event the company representatives had the opportunity to meet and discover common objectives. Since then several partnerships have been established. The WABE was able to establish a working relationship with the local public services. The operating authorities of the steelworks agreed to provide the electricity that is needed to run the RuhrtalFerry, which had been imple-

---

Success Story: Ruhrtal Camp

Inspired by the UK leadpartner the Artery partners from the Ruhr Valley initiated the first Ruhrtafel Circle. This gala dinner proved to be very fruitful. In order to strengthen the emerging synergies, the event was repeated. The representatives from the local companies were regularly invited to special locations for a stylish get-together. Since the establishment of the Ruhrtafel Circle, the contact between the public and private parties has grown closer. New public-private partnerships were set up that even went beyond the context of Artery.

In 2004 the Artery project partner WABE developed the idea of the Ruhrtal Camp, with the support of the Ruhr Valley Initiative. This summer adventure camp would be located on the shores of the Harkort Lake, one of the Ruhr's largest reservoirs. The camp would be a great recreation facility for children living in the region. In order to get such a camp off the ground, a lot had to be organised and provided. The costs for the participants would be incredibly high if everything had to be carried solely by the initiators.

The realisation of the first Ruhrtal Camp in 2005 was finally secured mainly through public-private partnerships with organisations participating in the Ruhrtafel Circle. Businesses from all branches of the private sector contributed to the success of the summer camp. The children received healthy meals every day, provided by the local catering and event company Schultenhof. The care institution Malterser Hilfsdienst took care of the water supply for sanitary facilities, and the local energy provider Mark E bore the costs of the installation and electricity usage. As the costs for the organisation and management of the camp were very much reduced, the fees for the children could be cut down to a reasonable price.

The innovative idea of creating a platform for the private sector, and thereby a beneficial situation attracting companies to public projects in the Ruhr Valley, was derived from the Artery partnership. Artery's working methods concerning public-private partnerships have become an accepted strategy in a region where public and private organisations have only little previous experience in working together. Since then public-private partnerships have been established in other development projects within the region.

Ways of Contacting Private Partners

~ **Through already established partnerships**
Advantage: you know who you are dealing with, as well as the company's priorities and objectives

~ **Through a politician/community leader**
Advantage: involvement of public representatives always increases the project's credibility

~ **Through publicity**
Advantage: partners identify themselves

~ **Through unconventional means: Be creativ**
Advantage: where no contacts to local businesses have been made yet, there is still a vast potential to exhaust!

mented as an integral part of the Artery pilot Barrier-free Tourist Attractions in the Ruhr Valley.

Creative, unconventional means have the added effect that they attract publicity. Through such publicity interested private partners are motivated to come forth with their own ideas on how they can become involved in the project. Such partnerships often contribute to the project in unexpected ways.

Motivating the Potential Private Partner

As mentioned before the creation of a win-win situation in a public-private partnership is essential. It is therefore important to point out the possible benefits of strong collaboration between the individual private partners. No enterprise will support a project if it does not see a potential gain. The experience of the Mersey Basin Campaign (MBC) has shown that it often proves useful to build a strong business case of what can be offered to the private sector. This clarifies the requirements and benefits both sides. Public and private partners know what they can expect, which is a good foundation for building trust in a working relationship.

However, within the context of the Artery pilot projects, main categories of incentives were identified. Knowing about these different kinds of benefits, a systematic search for private partners can be set up. Knowing which incentives a river landscape regeneration project can offer to a specific private partner, as well as knowing which benefits are most valuable for him, significantly increases the efficiency of the partner search.

Economic Benefits

In most cases it is financial benefits that motivate private companies to support a project. Clear monetary win-win situations are usually the case in public-private partnerships established for infrastructure related projects or for the construction of public buildings. Clear-cut definitions of the partners' tasks, as well as his risks and responsibilities are agreed on in the contract. They are usually of such a nature that it is within the partners' abilities to manage them.

If a construction company builds and operates a site, it will gain profit through future users. The company will ensure that the building is well constructed to avoid higher maintenance costs. If a bridge appears dilapidated, people will stop using it, while if it is well maintained, they are happy to pay a fee provided it gives them a faster route to their destination. In order to keep the financial benefit as high as possible, the operating partner will work as efficiently as possible.

In riverside development projects such financial profits for the private partner are mainly generated in the recreation, tourist and real estate businesses. In the Artery pilots one private institution that was convinced to support the project by financial gain, was the Yacht Club Harkortsee (YCH). In providing its land for the establishment of an information and service centre for tourism and local recreation in the Ruhr Valley, it participated in a public-private partnership with WABE and the City of Hagen. This service centre 'RuhrtalService' is an integral part of the Artery pilot project facilitating barrier-free tourist attractions along the Ruhr.

Within the scope of this project, the Yacht Club Harkortsee (YCH) broadened its services for disabled sports and water sports for schools. The restoration of the sailing club facilities and the revaluation of the whole terrain was the main incentive for the YCH to participate in the Artery project. Along with this came the perspective that the new scope of activities offered would attract new members. However, to sustain the enhanced facilities, it is in the Yacht Club's hands to maintain the site and the range of leisure activities, in order to sustain the profit gained through new memberships.

At the Speke and Garston Coastal Reserve, the prospect of financial benefits was also a great incentive. The expectation that improved riversides generate higher land prices motivated the property developer Peel Holdings to support the development of the new Coastal Reserve. It is already apparent that Peel's strategy has paid off. The regeneration of the derelict land gains Peel increased rental incomes from companies leasing land in the adjacent business park.

In riverside regeneration and 'green' urban development, monetary benefits are often rather subtle. Some-

times they set in only after the completion of the project. It is therefore important to make potential partners aware of this possible delay.

In the Ruhr Valley the owner of a housing company (and attendee of the Ruhrtafel Circle gala dinner), saw the Ruhr Valley Initiative's potential for the future of the whole region. The company decided to sponsor parts of different regeneration projects reasoning that if the region became more attractive, it would attract people and businesses to locate there. These people would need houses to live in, which the housing company could then let or sell. The private partner would thereby gain profit through the realised projects, on a long-term basis.

Improved Public Perception

A second important incentive is the improved public perception that comes with being recognised for being involved in publicly valuable projects. Potential partners have to be made aware of the difference

---

Opening ceremony of the RuhrtalService facilities

high quality offices overlooking a **sea of green**

Standing next to the Speke and Garston Coastal Reserve with the River Mersey beyond, this is a truly unique location offering a superb working environment that's perfect for attracting the top quality staff your business needs. Within that location you have the opportunity to create your own tailor-made facility. Whether it's a headquarters building, a service centre or just quality offices, the park offers leading edge accommodation from 10,000 to 500,000 sq ft (929 - 46,450m²) with generous parking. Wherever you do business, you're ideally placed here. In addition to the area's exceptional physical communications you're also served by massive broadband capacity locally and leading-edge GRID technology – companies here can trade internationally 24 hours a day.

the project will make to local people. This is especially interesting for companies that need to improve their public image, as it offers them the opportunity to change their public perception.

In the Artery programme, a participation in the Mersey Vale Nature Park in Stockport offered the water service provider United Utilities the opportunity to enhance its public position. The company operates its sewage treatment works close to the Heaton Mersey community. Evaporated gases from these sewage treatments produce strong odour problems. The bad smell is a permanent issue affecting the residents' quality of life. This has created a poor image for the company.

By working with Artery, in helping to maintain an important wildlife area on the sludge beds, United Utilities had an opportunity to build trust within the local community. Its involvement in a socially and ecologically important project demonstrates that the company cares about its neighbourhood. It shows that United Utilities is determined to contribute to the region's prosperity despite the problems originating from its industrial plants.

The energy supplier EnBW reasoned similarly concerning its power plant in Altbach near Stuttgart. The enterprise regards its relationship to the neighbourhood as very important and is therefore interested in keeping a positive image. It has sponsored the Heinrich-Maier-Park, which is located on former industrial land owned by the power station. The envisioned fish pass would enhance this park and contribute to the natural regeneration of the Neckar.

The energy provider established social and ecological responsibility targets to underline the corporate philosophy, apart from its economic interests. The involvement in Artery helps them to meet these targets. By working in partnership with the pilot project Wernau Neckar-River Reserve, the power plant is putting across the image of an environmentally conscious company.

Private partner Peel Holdings advertising the business park adjacent to the Coastal Reserve

Project Perception by Companies and Local Population

~ As the positive perception by the public is greater for projects with an obvious public value, private support for high-profile projects is easier found than for smaller ones.

~ It is therefore important to outline the value of the individual projects to the public, as well as to the potential private partners. If the project is perceived by the public as important, it increases the positive image of those who are involved.

Meeting Legal Obligations

Unlike the power station in Altbach, who voluntarily established corporate social responsibility goals, for other companies it is compulsory to meet certain legal obligations. Such obligations can often be made use of when motivating private partners to support socially valuable projects. Pointing out the opportunity for companies to meet their social responsibilities can be the trigger for some enterprises to get involved.

In one case Artery was able to make use of such official directions. The Dutch partners working on the Windlust project were able to win the housing corporation Woningstichting Ons Huis (Housing Corporation 'Our House'), because in the Netherlands such corporations have a legal obligation to provide affordable housing and participate in projects that benefit society.

The Housing Corporation 'Our House' was able to meet these legal targets by contributing to the reconstruction of the windmill. It offered valuable organisational assistance to the Projectteam Hollandsche IJssel relating to tax issues and construction co-ordination. The construction company also carried most of the project costs, especially in the construction of the windmill activity day-centre, which is now used as

a sheltered workshop by the care institution Algemene Stichting Voor Zorg en dienstverlening ASVZ (General Foundation for Care Services).

Positive External Effects

Urban development projects often have positive external effects on parties not directly involved in the project implementation. Such anticipated outcomes can be used by the project initiators to promote their entire undertaking.

Following this strategy, the Artery partners examined which external effects the implemented project might have on other organisations. Consequently they contacted the respective companies to promote these benefits. Even though a public-private partnership could not be established in every case, valuable contacts were made which might prove useful in future.

For the regeneration of the Speke and Garston Coastal Reserve, the UK leadpartner invited the National Trust into the project's Steering Group. The environmental and heritage charity is the owner of Speke Hall, which is adjacent to the Artery pilot. Being sensitive to the many historical characteristics of the site, the National Trust was keen to ensure that the Coastal Reserve was restored to its 'former glory'. Striving for a high-quality environment for the cultural heritage site of Speke Hall, they were a regular participant in the Coastal Reserve project meetings after the initial contact.

In the second Artery project in England, the Mersey Basin Campaign (MBC) followed a similar strategy. In its search for private partners for the Mersey Vale Nature Park, it contacted local businesses, which would profit from the large riverside space. The company Innov8 Technology was motivated to support the pilot project, because some of their staff used the community park to get to work, or for spending their lunch break. The nature park was promoted to Innov8 with the grounds of improving the working environment of the staff and also by providing potential team building exercises, for example tree planting activities.

Publicity

Another incentive for public-private partnerships is the positive publicity connected to high-profile public projects. The projects and also the companies involved gain recognition. The attention of the public

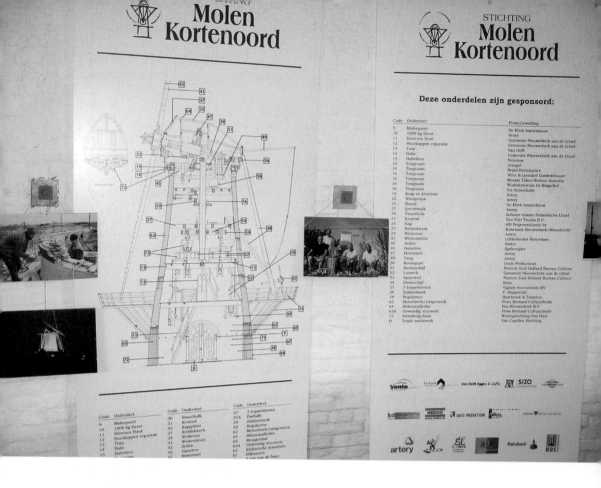

and the press increases the extent of how well a business is known in the region.

In order to promote the public and private partners of the windmill, the Dutch Artery team placed flags flying the different sponsors' names at the Windlust. Due to the prominent location of the windmill the flags can be seen by a large number of people. This provided excellent publicity opportunities for local companies and was an important motivating factor in encouraging them to support the windmill restoration. Among the sponsors who had their names on flags at the Windlust were the housing company Vestia and the care institution ASVZ, whose clients now work in the building adjacent to the windmill.

As part of the Artery project Barrier-free Tourist Attractions, the service RuhrtalVelo hires out special bikes for people who are disabled and gives information about suitable cycling routes along the river. In order to draw attention to these new facilities, the Artery partner WABE hosted an exhibition introducing these special bikes. Here visitors could gather information about the RuhrtalVelo and try out the differ-

ent bikes. This was only possible, because the special bike companies Trimobil and Judimed put several bikes at the disposal of the exhibition.

The event successfully promoted the new services offered in the Ruhr Valley, as well as the companies that build the special bikes. The companies gained new contacts to disabled associations and raised their profile in the region.

There are many motivating factors for private partners to support regional development projects, some diverse, others complimentary. It is therefore usually not just one benefit that encourages a private business to join a public authority within a working partnership, but often a combination of all those mentioned above. When initiating a project, it is therefore important to identify all possible benefits, and point them out to the potential co-operation partners.

Contributions from the Private Sector to Urban Development Projects

When trying to facilitate a project by establishing public-private partnerships, it is important to consid-

Main Incentives
for Private Partners

~ Economic benefits

~ Improved public perception

~ Meeting legal obligations

~ Positive External Effects

~ Publicity

To promote the different sponsors the Dutch Artery partners display detailed maps of the contributions inside the Windlust in Nieuwerkerk // Special bikes are the main attraction of the RuhrtalVelo service

er the nature of the individual endeavour. In regional development projects, the private sector is able to contribute in various ways. In Artery, private organisations and businesses have contributed to their local pilot project in different and sometimes unique ways. In some cases, the partnerships last much longer than the actual project implementation.

Financial Contribution

Project implementations usually require large investments. In order to reduce these costs, it is favourable to motivate private partners to contribute to the project financially. This monetary involvement can either be for a specific purpose or to generally support the development. Knowing how their financial contributions are being used makes the investor feel more involved. This psychological effect creates greater affinity for the project. In the search for sponsors and partners, it is therefore beneficial to either let the supporter know from the start what his money will finance, or even give him the opportunity to choose the purpose of his funding.

Basic Rules to initiate a successful Public-Private Partnerships

~ Work up a strong business case! Take time to sit together to look at the project and brainstorm opportunities. Looking at a familiar project from a different view point can prove to be useful!

~ Break the project down into smaller manageable pieces. This makes it easier to identify potential partners and mutual benefits. This process will also identify elements which are more and which are suitable to a public-private partnership.

~ Produce a potential marketing and promotion toolkit. Establish a 'catalogue of benefits' for private and public collaborations.

~ Find the appropriate person to make the approach.

~ Build trust! Be reliable!

For the Artery project Windlust, a monetary sponsorship was received for particular parts, as well as the overall reconstruction of the windmill and the connecting buildings. 25 local businesses came up with the costs for special elements, such as the front door and the sails. The companies were actively offered certain parts of the windmill. The names of the sponsors and their contributions are prominently listed on the project's website (www.kortenoord.nl). Other organisations symbolically sponsored a brick for the windmill wall, and in return had their company name engraved on it.

Through the private sponsorship of the construction elements and the bricks, a total of more than 100,000 euros was raised. These means made it possible to realise the reconstruction of the windmill to the aspired extent. The additional option to contribute smaller amounts, by taking on the sponsorship of individual construction parts or a symbolical contribution, made it possible for smaller companies, organisations and individual people to participate in a public project. This kind of sponsorship makes the partners' financial contribution to the reconstruction of a cultural heritage site visible. Furthermore, it creates a sense of responsibility for the future preservation of the windmill. The restoration of the Windlust has shown that often more partnerships and resources can be gained through many smaller sponsors, rather than focussing solely on major investors.

Peel Holdings' engagement of more than 100,000 euros in the Speke and Garston Coastal Reserve shows that financial investments do not necessarily have to be solely directed at material sponsorship. In order to support the implementation progress, the property developer financed all professional costs, such as legal fees, relating to the Reserve. Peel also engaged external lawyers to prepare papers to support the creation of a Coastal Reserve Management Company that will take care of the Reserve's future maintenance.

Service sponsorship might not provide a directly visible result, such as a construction element, but they both lead to the same result. The funding accelerates or facilitates the realisation of the project. Private partners, who attach great importance to sponsoring something 'visible' can nearly always be convinced by pointing out that the saved resources are often allocated to material requirements. In the end they are contributing to a visible result.

In-kind Contributions

A second method to use, when systematically searching for support for a public project, is to find potential partners respective to their material resources. Besides monetary investment, in many cases the private sector can assist with in-kind contributions. In public-private partnerships where the private party supplies mainly in-kind contributions, typical investments are building materials or operational equipment. However, depending on the nature of the aspired facility, in-kind contributions can be very diverse.

Possible Private Partner Contributions

~ Financial contributions

~ In-kind contributions

~ Expertise and manpower

~ Land agreements

In the broad Barrier-free Tourist Attractions project on the Ruhr, the bike retailer Trimobil agreed special rates for the purchase of the bikes for the Artery pilot. This enabled the RuhrtalVelo service to acquire higher specification bikes within the determined budget, allowing even more disabled people to use them. Trimobil has also agreed cheaper rental rates for larger groups. This is a great service for disabled sport groups, which like to utilise the special bike service along the Ruhr.

In the same Artery pilot, a second and rather different public-private partnership was established with the regional daily newspaper Westdeutsche Allgemeine Zeitung (WAZ). Newspapers reach thousands of people every day, and are an excellent medium to promote regional development projects to the public. It was a great success to win the WAZ as sponsor and multiplier for the Artery projects in the Ruhr Valley. Advertising in magazines and newspapers is usually very expensive. Therefore, printing the advertisements for the Ruhr Valley projects free of charge was an important contribution by the WAZ. The short-term benefit to the newspaper was the interesting content that it could offer its special readership. The major long-term benefit is the positive development of the region.

Bricks with names of sponsors on them. Many used the opportunity to take on a small sponsorship

Financial Contribution does not mean selling oneself!

~ When private partners contribute financially to a public project, it is important that they know they are not 'buying' the project!

~ The domination of the implementation procedures by one private partner can offend other project participants and in the worst case lead to their withdrawal and the failure of the whole project.

~ Clearly formulated contracts prevent unjustified claims or demands!

~ The project management and decision making process should stay largely with the project leaders and not be completely transferred to a private partner.

~ The project leaders should prevent becoming dependent on one financial benefactor.

These two established partnerships in the Barrier-free Tourist Attractions pilot demonstrate the different contribution possibilities to one and the same project. When involved in a public project, it is therefore important to think about all kinds of in-kind contributions, as not all good opportunities are obvious ones.

Sharing Expertise and Manpower

The third possible private sector contribution to an urban riverside development project is the sharing of expertise and manpower. This is one of the most active forms of involvement that a private partner can offer and again, includes all possible branches and organisation sizes, down to the commitment of a single private individual.

If an anticipated project requires the specialised knowledge of a certain subject, it is often possible to draw on a private partner's experience and resources. For the re-naturalisation of the Mersey Vale Nature Park, the Mersey Basin Campaign (MBC) was able to turn to its partnership with United Utilities and benefit from the company's habitat restoration expertise. Parts of the park's location are former industrial sites with grounds that require special treatment, in order to reintroduce local flora and fauna. As this is not the first regeneration project for United Utilities, the water treatment company has a vast knowledge and experience of restoring wildlife habitats on non-operational land associated with water treatment works. The company's expertise was shared and demonstrated to the pilot partners through a series of site visits to similar schemes in England's Northwest. It was therefore possible to apply the gained knowledge to the former sludge bed sites in the Mersey Vale Nature Park effectively restoring them to valuable wildlife habitats. United Utilities has also agreed to continue to offer advice on the management of these sites to ensure future maintenance.

Where some projects require expert knowledge, others need more practical assistance. Here companies can contribute by delivering services or taking on certain fields of work. Depending on the nature of the project, it is also possible for private individuals to participate.

This was the case with the maintenance works required in the Artery pilot River Bath Ruhr. When the open-air swimming pool was handed over to the operating society, the help of local clubs, businesses and also private individuals was needed, in order to keep operating, as well as to start the renovation and maintenance of the site. Members of local clubs like the Kanu Club Wetter (Canoe Club Wetter), the Deutsche Lebens-Rettungs-Gesellschaft e.V. DLRG (German Life Saving Association) and the relief organisation Technisches Hilfswerk THW (German Federal Agency for Technical Relief), supported the River Bath by running the technical operations. Others, including private individuals, helped out with the cleaning of the facilities, or by working shifts at the till during pool opening hours. Reducing manpower costs was necessary, in order to keep the swimming pool open. Sustaining

the swimming pool was a high priority among the local population, which is why different people with different skills came together to help out.

Contributing with expertise or manpower has a special benefit for the private partner. They know that their involvement is valued, because they contribute to the project in a very active and 'personal' way. It was their particular services that were needed and which integrated them in the implementation of the project. In such public-private partnerships, trust and a sense of responsibility are often much greater than in other kinds of collaboration.

Land Agreements – Securing Land for Public Use

In the UK many riversides are privately owned and not accessible to the public. In the Netherlands and in Germany, post-industrial brownfields also often belong to companies that have either relocated or simply stopped using the sites. These 'no-mans lands' have deteriorated over the years and are usually out-of-bounds to the general public. Integrating such land into public redevelopment plans is only possible with the landowner's consent. Therefore, securing private land can be essential for the implementation of riverside development projects.

The public-private partnership model is in these cases an excellent strategy to come to an agreement with private landowners. As it is mostly neglected and out-of-use land, the respective owners are usually interested in some kind of collaboration, especially if they see benefits on their side.

Land agreements can take many forms, from access permissions over leasehold agreements, to the donation or selling of land to create new ownership. For the Speke and Garston Coastal Reserve, landowner Peel leased out the whole terrain of the nature park to the newly formed Speke and Garston Coastal Reserve Management Company. As a special incentive for the management company, the lease for the Reserve is free of charge. The estimated value is one million euros. The benefits to the property developer have been explained above, but nevertheless,

Old river bath in Wetter waiting to be restored with private support

'Spin-off' – Benefits of Public-Private Partnership for Regional Development Projects

~ The contributions of the private sector often trigger 'spin-off' benefits that are not anticipated when the partnership is established.

→ The positive effects of public-private partnerships are often underestimated by the private sector. It is therefore important to actively seek these 'spin-off' effects and promote them to potential partners.

without Peel's consent the development of the valuable nature reserve for public use would not have been possible.

The possibility of gaining private land for public use through a collaboration between the public and the private sector is still underestimated. Selling or buying land is still considered to be the most plausible way of acquiring access to private land. However, many landowners are not prepared to sell their property. Valuable projects are often not undertaken, especially in riverside regeneration, as the option of establishing a public-private partnership with landowners has either not been considered or was shunned. The potential of establishing win-win situations through public-private partnerships should be more actively promoted in urban development and regeneration strategies.

Project Operation and Maintenance through Public-Private Partnerships

The most elaborate forms of public-private partnership are established to guarantee the operation and maintenance of a project. As they aim to ensure the quality and sustainability of facilities, these co-operations are never about single contributions. They are set up for a longer period of time and demand the constant commitment of the participants. In such long-term co-operations, the involved partners usually develop intense and trusting working relationships, from which both the partners and the project can only benefit.

In Artery, the partners were aware that riverside regeneration is only sensible if sustainability can be guaranteed. With public-private partnerships they aimed to ensure that the new public sites would not again fall prey to neglect and decay. These partnerships were targeted at the long-term maintenance and sustainability of the improved sites. Their different legal and organisational forms can be divided into four categories.

Maintenance through Individual Private Partners

One way of sustaining a project is to find responsible partners, who take care of all or certain fields of upcoming work. This can either be private companies that have special maintenance abilities, like landscaping resources or cleaning facilities, and are therefore interested in getting involved in the project, or companies that are already in some way connected to the undertaking.

The latter is the case in the Artery Vuykyard project, which undertook the regeneration of the former shipyard. The restaurant 'Fuik's Eten & Drinken', which is located on the newly developed waterfront community park, agreed to take on responsibility for the maintenance of the children's playground. The partnership between the restaurant operator and the municipality Capelle aan den IJssel, the local project partner, holds profits for both sides. The maintenance work carried out by the restaurant significantly reduce the costs for the public partner. A clean and safe environment for children attracts more locals to use the park. More people in the park mean more customers coming to the restaurant, and therefore more revenue for the company. The public and the private side supplement each other. When implementing a project for public use, it is important to consider business opportunities along the way and point these out to potential private partners.

In the case of maintenance through one or more public-private partners, as in the Vuykyard project, the

Hier wordt binnenkort g...

Grand Café FUIK'S Eten
Brasserie Drinken

PANNENKOEKEN - LUNCH - DINER - TERRAS

Opdrachtgever VILUSTO Vastgoed BV

Architect M III Architecten Rijswijk

ownership and main risks stay with the project initiators. This has the benefit for the public partner, that he reduces his costs without transferring any decision-making powers regarding the operation of the project. The private partner gains benefits, but bears only a small share of responsibility.

Operation and Maintenance
through a Multiple Ownership

Maintenance and sustainability of a site can also be established by transferring parts of the ownership. In this model the greater share of responsibilities and decision-making power regarding operation and management, are passed on to one or more private partners.

This model of a multiple ownership was realised in the Artery pilot project Windlust. After the successful completion of the restoration, the project initiators and owners, the Molen Kortenoord windmill foundation, sold most of its ownership to the housing corporation Woningstichting Ons Huis.

By keeping only a small share, the Molen Kortenoord foundation on the one hand diminished its own risks

and responsibilities concerning the operation and maintenance of the site, on the other hand, its remaining involvement gives them a say in the operation of the windmill. Since buying the main share, Woningstichting Ons Huis largely carries the responsibility for maintaining the restored and newly constructed buildings.

To reduce its own risks the housing corporation leased the neighbouring buildings to the care institution Algemene Stichting Voor Zorg en dienstverlening (ASVZ). The rental income covers a large part of the maintenance costs. Additional income for maintenance is raised through various commercial activities, like the selling of the freshly ground flour or private fund-raising.

In this way the Molen Kortenoord foundation accomplished excellent financial stability in the maintenance and sustainability of the windmill facilities, far beyond the Artery programme. This model of using public-

The private investor 'Fuik's Eten & Drinken' restaurant is supposed to increase the attraction of the Vuykyard community park

transforming riversides for the future

The Time and Bureaucracy Factor

~ When searching for private partners or applying for grants, possible supporters first have to be located and contacted.

~ Application formalities and contract policies have to become acquainted. Consider bureaucratic ways in larger corporations!

→ Normally both take time, and long fore-runs on the private, as well as on the public side! So start your search early!

~ Partnership development takes time! Whether you are working in a partnership with a contract or without any formal framework, consider that developing workflows, as well as establishing trust between the parties requires time.

ety the public body diminishes its risks without transferring the ownership. Instead the operating society bears the entire responsibility for the operation and maintenance of the project.

The City of Wetter, Artery project partner of the River Bath Ruhr, established a public-private partnership with the society 'Unser Freibad am See e.V.' (Our Open-Air Swimming Pool at the Lake Association), which had evolved out of a resident initiative determined to preserve the deteriorating local open-air swimming pool. In order to be able to take over the operation and maintenance of the pool, the local initiative turned itself into a legally registered society.

Without giving up ownership the City of Wetter transferred the risks of operation and maintenance of the old pool to the operating society, with the condition to reduce operating costs, as well as to renovate the dilapidated facilities. 'Our Open-Air Swimming Pool at the Lake Association' operates the pool until at least 2029. Until then the City's only responsibility towards the River Bath is an annual financial contribution to its maintenance.

In an operating society (a legally registered society), all profits gained are reinvested in the project. The private individuals, who run the registered society, are not liable with their personal property, if the society can no longer bear the operational risks and consequently cannot deliver. These were the main reasons the resident initiative chose this legal form to secure the operation of the swimming pool.

Establishing an operating society in the form of a registered society is a very promising approach to successfully sustaining a facility. A maintenance model, like the 'Our Open-Air Swimming Pool at the Lake Association', has great advantages for both the private and the public partner. While the public party reduces its maintenance responsibilities, it still owns the site and the facilities. The private partner however is able to make project decisions without being personally liable, if he can't carry on delivering.

private partnerships for operating and maintaining a public site has the great advantage that the public partner does not solely carry the main risks. The responsibilities are transferred and shared with the other partners. A possible shortcoming of this model can be that decision-making powers are also reduced through this partial privatisation. Clear arrangements are therefore necessary to prevent unnecessary disputes about competencies. The model of multiple ownership is a beneficial method of transferring risks, as long as the public side is prepared to share decision-making power as well as the project benefits.

Maintenance through an Operating Society

The third way to secure the long-term sustainability of a riverside development project is through the creation of an operating society. In an operating soci-

Maintenance through a Specially Founded Maintenance Company

Inspired by the way the Dutch partners sustained the maintenance of the windmill facilities, the UK Artery partners developed their own ideas on how to

Profit and Liability in a Registered Society

~ A registered society is a legal body.

~ Its members are mostly honorary staff.

~ All profit gained belongs to the society, not its members.

~ Should the society become insolvent, members are not liable with their personal property.

~ The chairman of the society is only personally liable if he commits an act of gross negligence.

guarantee the successful maintenance of the Speke and Garston Coastal Reserve.

In the beginning the project was led by a multi-agency steering group comprising representatives from Peel Holdings, the Mersey Basin Campaign (MBC), Liverpool City Council, National Trust, Mersey Waterfront, Randall Thorp and the Northwest Development Agency (NWDA).

However, a steering group is not an appropriate body for the management of publicly used private land. It does not have the legal or the human resources necessary for such an endeavour. In order to address this problem, landowner Peel and the MBC founded a special company: the Speke and Garston Coastal Reserve Management Company.

As two equal shareholders, the private partner Peel and the public organisation MBC, are the organisational and decision-making body of the company. They share all the risks and responsibilities for the sustainability of the Speke and Garston Coastal Reserve.

Patient of ASVZ; the care institution rented the facilities of the windlust

Three Key Factors for
Successful Public-Private Partnerships

~ **Commitment:** It is important that all participants in the working relationship support the project – public and private alike!

~ **Communication:** An intensive project-internal exchange among the partners is necessary to avoid misunderstandings. Public relations work is necessary, in order to inform the public about

the project, as well as the working partnerships. Transparency avoids public opposition to working methods or the project.

~ **Co-ordination:** All public and private partners in the project bear risks and responsibilities. To ensure a smooth operational workflow central co-ordination is indispensable.

Both, Peel and the MBC commit an equal fee to the management company, as well as supporting it with in-kind contributions and manpower.

In order to gain additional financial support for the further development of the Coastal Reserve, the man-

agement company founded an association for private sponsors. By joining this association, interested organisations can become private partners and contribute to the success of the local recreation facility and nature reserve.

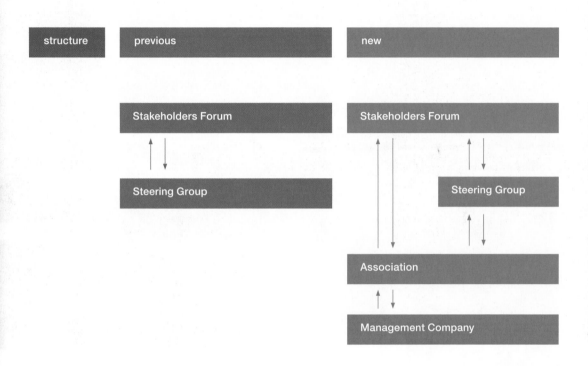

| structure | previous | new |

Stakeholders Forum

Steering Group

Stakeholders Forum

Steering Group

Association

Management Company

The association is not legally connected to the management company; however, its institutional goal is to support the development of the Coastal Reserve and the management company. For potential private partners, like the construction corporation Riverside Housing Company, this has the advantage that they are not liable, and therefore do not bear any risks regarding the reserve. In return for their contribution, which can be monetary, in-kind contributions or manpower, the members of the association have a direct consulting role in future development of the reserve. For the two shareholders, Peel Holdings and the Mersey Basin Campaign (MBC), this second public-private partnership with the association holds the benefit that they retain control over the project.

The combination of a management company and an association offers an interesting new method to involve additional private partners and facilitate further sponsorship opportunities. Unlike the former steering group, a company can engage further private partners and administer a maintenance fund. Another benefit is that a limited company is, unlike a charity, allowed to make profits. With the resources of the two organisations in the background, the management company can also handle all bureaucratic necessities, like enforcing insurances or leasing the land from Peel Holdings. All this is needed, in order to sustain the quality of a large terrain like the Coastal Reserve.

The Speke and Garston Coastal Reserve Management Company was established by the two shareholders for a time-frame of 25 years. Therefore, the initial private partners of the UK Artery project, MBC and Peel Holdings, are now in the unique position of being able to guarantee the quality of the Speke and Garston Coastal Reserve for at least a quarter of a century.

Headquarters of Peel Holdings; private partner of the Speke and Garston Coastal Reserve Management Company // Organisational structure of the Speke Garston Coastal Reserve Management Company

Lessons Learned in Artery

The numerous practical examples from the Artery pilot projects demonstrate how public-private partnerships can be used to facilitate the implementation and maintenance of riverside regeneration projects. However, due to their flexible nature, they are broadly applicable. Therefore, the working experiences gained by Artery offer plenty of insights for future projects.

To establish a Successful Public-Private Partnership the following aspects need to be considered

~ Identify possible opportunities to involve the private sector in your project.

~ Recognise the contributions the private sector can bring to the table.

~ Consider who could be suitable and who might have an interest in supporting the project.

~ Start to approach potential partners as early as possible.

~ Think about the different ways of contacting the individual organisations.

~ Make a business case! Promote your project actively!

~ Be sure to communicate all possible benefits to motivate potential businesses. Win-win situations are essential for a public-private partnership.

~ Urban development projects always have positive external effects. Make sure to internalise them!

~ Remember that public-private partnerships are flexible constructions. Build them according to the requirements of the project, but do not forget that companies have their own agendas.

~ Almost no project can do without future operation or maintenance. Consider in which ways public-private partnerships can ensure the sustainability of the project. The organisational forms of such partnerships can be very different. Be creative and create legal forms that suit the project!

Coastal Reserve and business park before regeneration: a specially founded management company is now taking care of the development of the site

Text: Frank Bothmann

Regional Strategies

Hollandsche IJssel: recreation and nature are important to the region's location factors

Introduction

Rivers and their landscapes are of exceptional importance to urban and regional development as they represent an attractive settlement area with a wide variety of economic, cultural and ecological potential. At the same time, river landscapes, by their very nature, do not stop at administrative borders: rivers flow through municipal, provincial and national boundaries. River landscapes therefore do not cover established formal planning regions. Consequently, informal procedures and approaches are therefore a requirement for regional strategic and development concepts that are geared to river landscapes. As they are informal, they are not subject to any formal guidelines regarding their organisational structure. As a result of this, the many different bodies seeking to develop their river landscapes in the regional context to a sustainable degree, are faced with the difficult task of finding an effective and sustainable approach. For this reason, the aim of the Artery project has, from the very beginning, been to cull a variety of methods and experiences from the regional development of river landscapes, to exchange their know-how and to come up with success factors for designing a regional development strategy for river landscapes. Since the design significantly depends on the specific regional conditions and development prospects, there can be no such thing as a generally applicable strategy for the regional development of river landscapes.

Regional Development Strategies

Regional development strategies are long-term models for a given region. They create guidelines for the future development of a region. They define common goals for a variety of fields of action, including their implementation. They should be geared to action and, on the strength of these concrete goals, should foster measures and projects. Regional development strategies provide a framework for co-operation that simplifies or even facilitates how the various bodies and institutions work together at a higher spatial level.

Given the increase in globalisation during the last century, the need for regional development strategies has risen enormously. Production factors such as labour and capital are very flexible in today's economy. The rise in mobility and new technical developments make it possible to shift production factors quickly. Administrative borders pose virtually no cultural, geographical or legal barriers. As a result, competition among the regions has increased, be it as residential, production, sales, recreational or investment locations. Regions are subject to an increasing amount of competition on a global, national and regional level. Demographic development also drains structurally weak regions of its inhabitants. For this reason regions must project an even greater positive image of themselves and highlight their specific additional value over the competition if they want to survive in the long term.

Regional Development Strategies

Regional development strategies are integrated concepts for the co-operative development of a region, covering several municipal boundaries. For the most part informal and non-binding in a legal sense, especially in the case of regional riverside development, the dialogue-orientated voluntary commitment of all involved leads to the development of long-term, action-orientated approaches for sustainable development.

European Spatial Development Perspective – ESDP

ESDP seeks to gradually achieve a sustainable and regionally balanced development within the territories of the European Union.

It pursues three main objectives::

~ Economic and social cohesion within the EU

~ Sustaining the natural resources and cultural heritage

~ more balanced competitiveness within the European Union

Regional Development Strategies for River Landscapes

A look back in time reveals that rivers have always played a major role in the development of a region. They have always been key components in the economic, ecological and social area. As competition between the regions increases, the prospects for river landscapes are particularly good. Rivers have become an important selling point, especially since the declared aim of creating greater cohesion of economic conditions in the European Union. The traditional determining location factors such as infrastructure, taxation, labour costs, levels of education among the potential labour force, sales potential and others, are increasingly being brought into line.

Within the scope of regional strategies, 'soft' location factors – in addition to the so-called hard factors – are therefore gaining more importance in assessing locational advantages for new and relocating businesses, institutions, residency or major events and attractions. Such factors include good environmental conditions, good water quality and also attractive local recreation as well as appealing cultural and leisure activities. The sustainable development of river landscapes especially enables these factors to be expanded with the aim of reaching sustainable economic, social and ecological development of the region.

Regional Development Strategies in the Artery Project

Examining the different approaches undertaken to develop river landscapes in the partner regions, was one of the core issues of the Artery partnership. A comparison of the different approaches taken by the five Artery regions revealed commonalities which emphasised the key aspects of a regional development strategy. The experience gained here also shows planners the creative and innovative means of implementing the European Spatial Development Perspective (ESDP), as well as the European Water Framework Directive.

Under the guidance of the Regionalverband Ruhr (Regional Association Ruhr, RVR) as the thematic leadpartner, the determining factors for successful river development strategies were studied. Together with its predecessor institutions, the RVR can look back on decade-long experience in regional planning and development and has also played a significant role in the implementation of a variety of regional strategies. Since being restructured in 1979, its main core

competency has been the regional development of open spaces. Examples of this are the Freiraumsystem Ruhrgebiet (Regional Open Space System in the Ruhr area), the Natur- und Freizeitverbund Niederrhein (Nature and Recreation Network Niederrhein) and the Emscher Landschaftspark (Emscher Landscape Park). Through its Ruhrtal Initiative, the RVR – together with other regional and municipal partners – created a new development strategy for a river landscape along the Ruhr.

Concept for a Regional River Landscape Development

An overall spatial development concept at regional level can provide the content framework for co-operative riverside development. It forms the basis for agreed objectives and fields of action among the players concerned. A concept expounds these for the future development of the region. This can facilitate co-ordination of the individual sub-projects and synchronisation of the overall project. Clear, spatial boundaries also prove useful as they clearly show which

areas, administrative bodies and other players are involved in this co-operation. In the process, however, it should be ensured that these boundaries are not too narrow as they might otherwise prove too restrictive for effective co-operation.

Aims and Target Groups

Regional riverside development concepts and their aims should be developed together with all co-operation partners involved and be based on the region's specific problems, potential and prospects for development. Depending on the initial situation, completely different courses of achieving sustainable riverside development may emerge.

An indication of the extent to which specific circumstances shape the development of regional strategies can be found when comparing the initial situations and the aims of the development concepts in the

View from the 'Kanzel' in Hattingen-Blankenstein on the Ruhr: a lush landscape in stark contrast to the prevalent perception of the Ruhr region. Integrated regional development supports this change of image

A Comparison of Initial Situations and Aims		
Region	**Initial Situation**	**Objective**
Ruhr Valley	· Economic structural change together with the social, ecological and economic challenges that come with this · Derelict areas · Loss of jobs · Potential for leisure and tourist activities · Poor image of the region	· Promote leisure and tourist activities · Enhance the regional image · Create jobs, new facilities and economic possibilities · Attract people and towns back to the river
Mersey Basin	· Economic structural change together with the social, ecological and economic challenges that come with this · Water pollution · Contaminated and fallow/derelict areas · Poor image of the region · Development of a regional park along the river	· Enhance the regional image · Sustainable development of the Mersey Basin · Foster economic development · Involve the inhabitants · Create range of local recreation activities · Improve water quality
Hollandsche IJssel	· Contaminated riverbanks · Implementation of a national plan to safeguard green areas in the region · Poor image of the region · Industrial heritage under threat	· Integrated development of the riverbanks · Restore the riverbanks and other contaminated areas · Create new leisure activities · Protect industrial heritage
Rhine-Neckar	· Development of a regional landscape park between the Palatine and the Odenwald Forest · Implementation of the land utilisation and landscape plan by means of the co-operation project 'Living Neckar'	· Integrated and sustainable development of the riverbanks and creation of a green axis for recreation activities and nature conservation · Build public awareness · Create leisure activities
Stuttgart-Neckar	Development of a regional landscape park which is characterised by · agriculture/forestry · an integrated habitat system for flora and fauna · an integrated system of ways to develop recreational areas and valuable farming areas	· Ecological flood control measures · Greater bio-diversity · Create range of local recreation facilities · Urban renewal combined with cohesive, accessible green areas along the Neckar

Artery partner regions. For all, the main aim is to enhance the attractiveness of the region.

In the Ruhr Valley and the Mersey Basin, structural change – combined with social, economic and ecological issues – plays an essential role in the development and renewal of riversides. In order to establish itself as a region for recreation and tourism, the Ruhr Valley seeks to improve its landscapes and expand the range of recreational facilities – with the river Ruhr being the unifying element. This development is intended to strengthen the factors supporting the Ruhr as an attractive residential and business location. The measures taken in the Mersey Basin also focus on the ecological renewal of the river and its banks with a view to enhancing the quality of the area to offer a more attractive surrounding for residential and economic purposes.

For both these regions as well as the Hollandsche IJssel, regenerating the respective river is intended to raise the perception of the region. The Hollandsche IJssel and its surroundings have long been conside-

red 'the backyard of the region' and the 'dirtiest river in Holland'. The integrated sanitation of the heavily polluted river and the contaminated soil on its banks restores the area to an attractive river landscape and provides an opportunity for this region to divest itself of this negative image.

The development of the riversides in the Stuttgart-Neckar and the Rhine-Neckar regions follows the idea of creating an attractive landscape for recreation. It is intended to enhance locational advantages and thereby attract people and businesses alike to settle in the region.

Beginning with the general aims for the future development of river landscapes, it is recommended that suitable target groups are selected in order to be able to act effectively and target-oriented. One of the main target groups of development strategies in the Artery partner regions are their respective inhabitants. Only in the Mersey Basin were the measures specifically geared towards the communities and the residents living directly on the riverbank. To achieve sustainable development, it is sensible to specifically address schools as well as educational institutions and facilities which have an interest in the river in order to raise lasting awareness as has been the case in the Rhine-Neckar region. Another key target group is the businesses situated in the region. They can be encouraged to increased ecological responsibility for the river, or even be motivated in becoming partners within public-private partnerships. This is actively being pursued by the Mersey Basin Campaign (MBC).

Planning Approach

A general planning approach can act as a theoretical superstructure for the envisaged objectives and builds a common ground within the development process. It includes a brief, generally comprehensible description of the regional development strategy and its goals. In doing so, they provide the criteria for selecting sub-projects and are also useful as a means of gaining political and public support. As a mission statement they also offer decision makers and the public a means of orientation in the complex process involved in developing river landscapes.

Fields of Action and Measures

The scope and nature of the regenerative measures depends on the specific circumstances of a region and the envisaged goals of the regional partnership. In order to lastingly improve soft location factors, it

Fields of Action and Measures

Deciding on the number of issues and fields of action as well as which players should be involved in drafting the concept for and implementing the regional strategy, is like walking on a tightrope. How many issues can be covered? Who should be integrated? Covering too many issues, incorporating too many bodies may often result in ineffectiveness. Covering too few issues and incorporating too few players leads to the initiative failing to have any sustainable effect.

is advisable to integrate measures geared towards enhancing regional identity and the quality of life of the inhabitants alongside strengthening the economy.

Along the Ruhr and the Mersey, the riverside redevelopment measures focus on the fields of recreation, leisure activities, tourism, cultural and industrial heritage, living and working on the river. These measures encourage cities to turn back to the river again and to benefit from the unique qualities of their riversides. Improving access to the river and opening up the surrounding countryside are therefore crucial design features with respect to the future users. Along the Mersey, these measures are being complemented by the urgently needed improvement in water quality.

In the case of Hollandsche IJssel, the sanitation of the contaminated areas is being undertaken with a clear focus on the needs of its future use. In part, it is leaving the original use untouched whilst creating new possibilities in other areas like nature, recreation, living and working.

When developing a regional strategy for riversides along the Neckar, the main focus is on raising public awareness and increased participation of the local population. In the Rhine-Neckar region, the development of the riverbank as a cohesive green axis is based on the fields of action pertaining to nature con-

Scope of Regeneration Measures of the Ruhrtal-Initiative	
Fields of Action	**Measures**
Experience the Ruhr Valley	· Historical RuhrtalBahn · Shipping on the Ruhr · The Ruhr water trail · Ruhrtal cycle path
Leisure and tourism in the Ruhr Valley	· Historical heritage and industrial culture · New leisure activities and ecology
Landscape and urban planning programme 'Cities on the Ruhr'	Urban planning and implementation
Regional marketing of the Ruhr Valley	Raising public awareness through PR measures and (cultural) events

servation and recreation. The measures undertaken by the Verband Region Stuttgart (Association Region Stuttgart) include the co-ordinated development of areas for settlement and transportation as well as the conservation of undeveloped areas for nature and recreation. The fields of action relate to ecological flood protection measures, the interconnection of biotopes, protecting the cultural landscape and restoration of an accessible riverbank.

Formal Planning Procedures

Given the nature of their co-operation, regional development strategies for river landscapes are usually informal. This having been said, they need formal planning instruments especially where this concerns the implementation of regeneration projects. It is imperative that these are embedded in formal planning procedures. The development of riversides can

Riversides have to cater to very different needs: here at the Hollandsche IJssel, commercial purposes meet residential neighbourhoods. Regional Strategies help to keep a balance between the different interests

benefit from existing formal plans and strategies as well as assist in their further implementation.

In the Artery partner regions, differences are certainly discernible and vary widely depending on the respective legal and spatial planning framework. Whilst some have formal regional planning, others have no binding primary plans or strategies. In the Stuttgart-Neckar and Rhine-Neckar regions, the development of riversides serves to implement existing regional and landscape plans. In the Mersey partner region, the spatial planning of the local authorities must also be aligned with binding primary development strategies. In the Hollandsche IJssel region, the situation is quite the reverse. The partners involved in the Project Hollandsche IJssel are bound to integrate the spatial goals of riverside development into their regional and land utilisation plans.

Planning Horizon

Co-operation at regional level in general and the development of riversides in particular are complex, multi-layered processes which, depending on the ini-

146

tial situations, require a correspondingly long planning horizon.

The action plan in the Project Hollandsche IJssel has been set for a period of 12 years, for example. The Ruhrtal Initiative has a planning horizon of 10 years while the Mersey Basin Campaign (MBC) assumes that it will take 25 to 30 to develop the riversides. In addition, the Artery partner regions are at different stages of development. Whilst the processes in the Stuttgart-Neckar and Rhine-Neckar regions are still in their inception, the other regions can already look back on many years of experience. The initiatives at the Hollandsche IJssel and in the Ruhr Valley were both launched in the late 1990s, whereas MBC began its work as far back as 1985.

Players and Organisation

To develop a region as a whole and to pool the varying interests along the river with a view to achieving sustainable development, regional co-operation and partnerships form the backbone of any river landscape development. Co-operation lives by the players involved in the process who assume joint responsibility for it.

Initiators

Regional strategies for riversides can be initiated at different levels. Such processes are frequently strated by several players who are then brought together to form a partnership at a later stage. Very often, these actors derive from local and regional administration. To initiate a regional strategy with success, it is often beneficial to gain support from a well-known personality. This person can bring home the idea of regional riverside development to other key decision makers and institutions.

Given the long planning horizon for riverside development, it is advisable to have a public institution as the initiator and organiser of the development process. A regional administration body in the position to make the necessary political and financial decision would also be expedient.

In the Ruhr Valley, municipalities the Ennepe-Ruhr County and the Regional Association Ruhr, founded the Ruhrtal Initiative and established a network aimed at developing common potential for the Ruhr

Valley. On the Hollandsche IJssel, the towns turned to the provincial government in the Province of South Holland to overcome the problem of contaminated soil along the river. The provincial government is legally responsible for soil decontamination schemes and application for the respective subsidies. In this specific instance and on the strength of its administrative tasks and its means of conducting action, the province was in a position to establish common regional goals for riverside development together with a total of 13 municipalities and institutions. In the Mersey Basin, the waterways had been so contaminated that fundamental measures became unavoidable. This process was initiated by a number of organisations and groups as well as by Britain's then environmental minister, Michael Heseltine. In the Rhine-Neckar region the landscape development of the Neckar between Heidelberg and Mannheim was initiated in the context of land utilisation and regional landscape planning.

In the Stuttgart region, the overriding regional players, such as the Association Region Stuttgart, are currently attempting to get an implementation-orientated landscape development strategy with a focus on riverside regeneration off the ground.

Groups of Players

Depending on the initial situation in the region, it may be advisable to include completely different groups of players. In accordance with the European Water Framework Directive, which is of major significance to the concept and implementation of riverside development, everyone with a vested interest in riverside development should be involved. To what extent this is expedient depends on how far the overall project has progressed and on the spatial nature of the respective measures. Local level is best suited for incorporating third parties as these are directly affected by their immediate surroundings. This level is also suited to ensure that the measures are geared to the target groups. In many instances the higher regional level appears to be too abstract to involve the general public.

To promote riverside development, it is beneficial to involve partners from the industrial and tourist sectors. Also water authorities, land owners, city and local councils, expert planners and potential sponsors

Regional Riverside Development and the Water Framework Directive

~ To conduct future riverside regeneration projects it will be important to heed the European Water Framework Directive (WFD). It came into effect in 2000 and calls for the 'good ecological condition' of all waters within the European Union. The Directive is the first of its kind to provide a holistic and integrative approach to sustainable water management along the entire river catchment area.

~ Sustainable riverside development can help to meet the goals of the WFD. In return, better water quality assists the regeneration of the riversides which, for its part, stimulates the economy and regional development. Good water and environmental quality are part and parcel of the most essential soft location factors for economic activity and decisions as to where to locate. The WFD can therefore act as an instrument to promote the regeneration of riversides and sustainable regional development.

should be involved from the outset. To win over these and other relevant players to develop a common concept, their interest in regenerating riversides must be sparked. This can for example be achieved through workshops and other special events. The prospect of receiving subsidies can also help to instigate regional co-operation.

Generally speaking, it is vital to highlight the benefits of regional co-operation to all those concerned. To gain political support for the development process, it is advantageous if results can be put forward at an early stage. The results of successful regional co-operation often meet with a positive reception among the general public and secure political support.

The groups of players incorporated into the development strategies of the Artery partner regions differ both in their degree of active involvement and in their decision-making powers. Operative responsibility for the concept and implementation of these strategies is, in part, highly divergent.

Administration and Politics

In the Ruhr Valley and the Hollandsche IJssel, the towns and municipalities are actively involved in the development measures. They also have voting powers at overall project level and are represented at control level through their political representatives. Since the implementation of local projects in the Ruhr Valley is the responsibility of the city councils, their implementation and funding are discussed in local parliaments.

The Project Hollandsche IJssel is one of the nine national projects in Holland's 'Green Heart', a preferential land use area for the maintenance of open green and nature conservation space in the area of the densely populated 'Randstad' conurbation. For this reason, the project is of significance in terms of structural development and has gained support at all levels.

In the Stuttgart-Neckar region, the Association Region Stuttgart is the initiator and a key actor in the implementation of regional riverside development. Here, the towns and municipalities are integrated at project level and are repesented by a regional politician in the regional council. All of the planned measures and concepts must be submitted to and passed by this body. A law which came into force in November 2004 now vests the Association with responsibility not only for the planning, but also the implementation of a regional landscape park where the Neckar-Landscape Park is a permanent component. As a body responsible for a measure, the Association Region Stuttgart can now take a direct active role in the implementation of the landscape park.

In the Mersey Basin, riverside development is supported by the British government and the government's North West office attends the meetings organised by the Mersey Basin Campaign (MBC) as an observer. At local level, the environment agency, in its capacity as

observer member, networks the local environmental activities with the activities undertaken by the MBC. Local representatives, in certain instances council members, but most notably employees of the towns and municipalities, are integrated locally into the individual River Valley Initiatives in the river catchment areas.

In the Rhine-Neckar region, the Nachbarschaftsverband Heidelberg-Mannheim (Neighbourhood Association Heidelberg-Mannheim) acts in co-operation with the Verband Region Rhein-Neckar (Association Region Rhine-Neckar) as the co-ordinator and promoter for riverside development. Responsibility for the implementation of the particular projects lies with the local authorities.

Residents and Stakeholders

At the Mersey, the Hollandsche IJssel and in the Rhine-Neckar region, the general public plays a major role in decisions relating to, and the implementation of measures. In every river of the Mersey catchment area, local residents' initiatives, charities and clubs,

among others, can participate in existing action partnerships. In the case of Hollandsche IJssel, a variety of stakeholders are not only involved in the local implementation of measures but also take on an advisory role in the planning process. In the Rhine-Neckar region, public awareness plays a crucial role in the sustainable implementation of riverside development. Here, schools and other educational institutions are involved in the development process. In the Ruhr Valley and in the Stuttgart-Neckar region, residents are addressed on special occasions and events.

Economy

In the Ruhrtal Initiative and the Project Hollandsche IJssel, representatives of local and regional industry also have an advisory function. They are also involved in the project implementation. At the Mersey, an association has even been founded which maintains

A picturesque scene along the Neckar: The landscape park between Heidelberg and Mannheim sensibly combines recreation with environmental protection

The right degree of political support

~ The development concept should include clear objectives and measures

~ The measures should quickly show tangible results

~ Politicians should be kept abreast of the progress and results of the development process on an ongoing basis. In this context, it is useful to work in close co-operation with the local press

~ Politicians should be actively involved, for example as patrons or at inaugurations, project presentations or events

~ Renowned politicians should be won over as initiators and supporters with a view to bringing home the idea of regional riverside development to high-ranking key actors

~ All measures should be nonpartisan so as to avert the possibility of political instrumentalisation or polarisation

contact to representatives of the regional economy and offers them the possibility of supporting regional development. Members of this association also have voting rights on decisions which can be taken at regional level. Through such active incorporation, they can be sensitised to river related ecology matters and in turn, create opportunities for supplementary public-private partnerships.

Organisational Structures

For the work to be successful, structures are necessary which regulate the communication channels, division of tasks, responsibilities and decision-making

powers. The partner regions have applied different organisation models in keeping with their legal and planning frameworks and their respective initial situation.

Since riverside development requires a long-term planning horizon, a prerequisite for the work to be successful is personnel and organisational continuity. Knowledge and personal contacts within a well-functioning network are difficult to transfer. For this reason staff changes can lead to a loss of information and impair the commitment for a co-operative process. A manageable group of people should therefore assume the process control and form an enduring core within the co-operation.

The Project Hollandsche IJssel, the Mersey Basin Campaign and the Ruhrtal Initiative each have a steering group which oversees the development process in their river region. Depending on the specific initial situation, however, the steering group has a different sphere of responsibility and different groups of bodies are incorporated.

In the case of the Ruhrtal Initiative, the steering group decides on fundamental issues and on which projects are to be included in a common action plan. The steering group comprises political representatives from local and regional authorities as well as representatives from the state ministries and the regional economy. As with the steering group in Holland, where representatives from the water boards are also involved, it has the task of securing political support. In the Project Hollandsche IJssel, the steering group also has responsibility for overall project co-ordination and maintains contact to the relevant stakeholders. It reaches programmatic decisions on the organisation, planning and implementation of measures and is responsible for reporting. The areas of responsibility for the steering group within the MBC are similar. Here, the Campaign Council reaches decisions on key strategic issues as well as an annual action plan. The Council comprises representatives from regional partners as well as representatives from the regional economy, the Voluntary Sector and North West University.

Whereas many of the operative tasks are assumed by the steering group in the Mersey and Hollandsche IJssel projects, these are the responsibility of a regi-

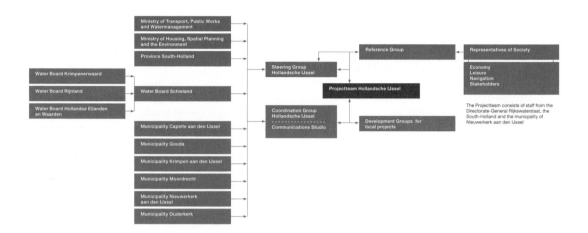

onal workgroup in the Ruhr. This Regional Working Committee consists of the heads of the building and planning departments of the project partners and their representatives. It translates the strategic guidelines of the steering group into concrete measures and monitors the implementation of the overall 'Ruhrtal' project. It advises and reaches decisions on the implementation of sub-projects and nominates project executors. In addition, it provides the information on which the decisions of steering group are based.

In the case of the Hollandsche IJssel, a co-ordination group assumes the task of steering the individual projects. In this group each authority involved in the projects is represented by a project co-ordinator. It marks the management level of the projects and is responsible for financial monitoring and the contextual connection between the parts as an overall project. With the help of the co-ordination group and through staffing the Projectteam Hollandsche IJssel - the executive body at overall project level - the steering group accompanies and supports the overall implementation of the project.

In both the Hollandsche IJssel and the Ruhr Valley regions, the various organisational units are also supported by external service providers who assume tasks such as administration, project co-ordination, public relations, fundraising and project implemen-

tation. In the case of MBC, these and similar tasks are assumed by so-called advisory groups. These advisory groups are automatically full members of the steering group. In the Hollandsche IJssel and Ruhr

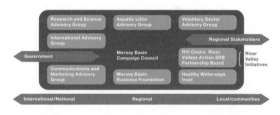

The organisation charts of the Artery partners at the Hollandsche IJssel (top), the Mersey (middle) and the Ruhr (bottom)

Communication Structure

Establish clear channels of communication and decision-making processes between the overall steering level and the operative levels. Ensure that every level here is afforded equal opportunity to express and represent their concerns. Both strengthen the commitment of those involved and this usually results in additional support. Furthermore, this can also lead to support from the political decision-making level being secured.

Success story:
Public Participation at Regional Level
The interests of the citizens are permanently represented in the organisational structure of the Project Hollandsche IJssel. Participation in the so-called Klankboard reference group ensures that the citizens have access to information for the project's lifetime and allows for their participation in key discussions. It marks the consistent implementation of the principle of public participation at strategic level.

The Reference Group
· comprises representatives from a broad range of stakeholders
· is chaired by a member of the steering group
· advises on and evaluates the agenda of the steering group
· can incorporate its own issues
· is consulted on fundamental issues
· has no voting rights

regions, these experts are only consulted as and when needed.

In the Hollandsche IJssel project, a wide variety of citizens' initiatives and stakeholders are pooled in a reference group. They have an advisory role and provide their comments to the steering group. This reference group can also develop and suggest own ideas. It is a permanent feature of the organisational structure of the Project Hollandsche IJssel.

In addition to these permanent organisational structures, MBC and the Ruhrtal Initiative organise workshops, competitions and campaigns involving a wide range of private and public parties in order to generate further project ideas.

Organisational Forms
In principle, regional strategies work on the assumption of close co-operation and partnerships between the various parties. Who should initiate the process? Which partner should be won over for these networks? Who should and can be involved? How should the network be organised? What legal framework is required? Numerous possibilities are available when forming these important network partnerships – ranging from loose agreements to written declarations of intent and communiqués, all the way down to contracts. Where possible, these issues should be clarified in the initial phase of the development strategy.

The implementation of the development strategies occurs on a voluntary basis in the Artery partner regions. Only those partners involved in the Project Hollandsche IJssel are tied to a contractual implementation agreement through which they are committed to assuming responsibility for attaining the common goals. Among other things, this agreement governs the organisational structure and the distribution of funds. The 'werkboek' action plan is a dynamic part of the agreement. Over the course of the project's implementation it is adjusted in line with the progress achieved and specifies a person in charge for each individual project.

Under construction: the canoe access steps to the Mersey at Stockport. Better access to the river allows for new opportunities, an important element in sustainable riverside development

Success story: River Valley Initiatives The development of the riverside in the Mersey Basin also includes its numerous tributaries. To achieve its main objectives, the Mersey Basin Campaign pursues so-called 'River Valley Initiatives' (RVI). These 19 individual initiatives form in many ways the operational core of the Campaign. They provide strategic leadership on a local level and directly help to restore the quality of the waters, to regenerate the riverbanks and to raise public awareness. In this respect, they play a vital role in implementing the regional strategy.

The River Valley Initiatives contribute in the following areas:
· Quality of the river basin: each RVI has strict guidelines on the disposal of waste and organises clean-up activities at regular intervals
· Riverbank regeneration: RVIs assist the larger projects in the region with their planning, funding and citizen involvement
· Involvement of communities, individuals and businesses:
The RVIs work at local level with schools and companies, accompany projects run by students; write articles for the local press, newsletters and the MBC website and organise walking tours

Implementation and Funding
Regional Development Phases

The development of riversides is a long-term and complex process. To be successful, the process needs to have a programmatic structure. Even if it is not possible to have one generally valid regional strategy due to the various framework conditions that prevail, there is an ideal type of approach towards developing such a strategy. This overview structures the regional strategy concept according to its various elements. It does not represent a binding chronological order but, where various different approaches exist, it serves instead as a means of orientation. In order to remain flexible throughout the planning phase

and to be able to react in the appropriate manner to any changes in the framework conditions, the individual phases should flow into one another.

The general stages of preparation, concept and implementation of a regional development strategy are divided into five basic phases. Ahead of this is the general mission statement of the planning approach. As the perspective for the future, it initiates the first phase of finding a concrete strategy.

Preparatory Phase

In Phase 1, the existing situation of a given region should be analysed using a variety of methods such as statistical studies, opinion polls and SWOT analy-

Phases of a Regional Development Strategy

~ Preparatory Phase
Phase 1: Stock-taking and analysis
Phase 2: Development of general objectives and establishment of organisational structures

~ Concept Phase
Phase 3: Selection of fields of action and determination of target groups
Phase 4: Development of concrete measures and projects

~ Implementation Phase
Phase 5: Funding and realisation of projects; ongoing conceptual adjustments and results control

ses (strengths, weaknesses, opportunities, threats). At the same time, all of the key actors should be brought together to draft the next steps to be taken. From this, Phase 2 should develop a vision of the future for the region and goals should be formulated. Following on from this, a strategy should be developed for realising the vision.

In doing so, a so-called bottom up approach is considered vital and indispensable for achieving successful regional development. In this case, the regional development concept is drafted through the joint efforts of all those affected. This raises the level of identification with the overall project and supports the implementation of the local and regional sub-projects.

On the Hollandsche IJssel, the preparatory phase largely comprised of organising the initiating partners by forming a steering group and of conducting an analysis of the actual situation complemented by a cost calculation. The possible future use of the sanitated riverbanks was identified by means of a study. In the

Ruhr Valley, the initiating parties had already worked together as part of an application process relating to a regional funding programme on spatial development provided by the state of North Rhine-Westphalia. The first phase began by further developing the concept which had been drafted at that time. In doing so, a process was initiated to generate visions and development ideas for the region. The result of this process was a memorandum of understanding which reflected the first signs of the Ruhr Valley strategy and its development concept. Subsequently the first feasibility studies were commissioned.

Lead by the cities Heidelberg and Mannheim a total of 18 municipalities from the Rhine-Neckar conurbation have joined forces in the Nachbarschaftsverband Heidelberg-Mannheim (Neighbourhood Association Heidelberg-Mannheim) to develop a joint land utilisation and regional landscape plan. As responsible planning authority the Neighbourhood Association developed the regional landscape project 'Living Neckar' in 1996.

In the Stuttgart-Neckar region, this initial phase saw the development of a concept for a regional landscape park. It is based on an analysis of the open space areas in the region. As part of this project, the Association Region Stuttgart developed the idea of a Neckar-Landscape Park. In the step that followed, the concept of a Neckar-Landscape Park was put forward and discussed with the relevant stakeholders. It was then agreed that the next developments would be accompanied by a steering group comprising representatives taken from these stakeholders.

Concept Phase

The concept phase complements the preparatory phase. Using the findings of the stocktaking and analysis, based on the general objectives, Phases 3 and 4 deal with selecting the fields of action and determining the target groups. On the strength of this knowledge, issue-specific measures are then developed. At the same time, an organisational structure with the necessary capabilities should be set up. During the concept phase, the Hollandsche IJssel and Ruhr Valley regions developed a regional action plan. In the Ruhr Valley, developing the tourism sector was identified as an important objective and field of action. The

Neighbourhood Association Heidelberg-Mannheim designed the project 'Living Neckar' as an open planning procedure for all parties involved including municipalities, technical authorities and associations as well as participation of the general public. In the Stuttgart-Neckar region, a variety of project concepts were drafted at local level.

Implementation Phase

The concluding Phase 5 contains the necessary measures for sustainable funding and realisation of the individual projects. As part of an integrated regional development their implementation is also reviewed and evaluated in this stage. In the case of Hollandsche IJssel and Ruhr Valley, this phase is characterised by the parallelism of the idea, planning and implementation phases. While the implementation of the individual projects in these regions did not start until a regional development concept had been developed, the implementation of the projects in Stuttgart-Neckar and Rhine-Neckar, triggered enough motivation to effectively establish a lasting approach to riverside

development in these regions. The control and review mechanisms installed here are designed to monitor and evaluate the strategy and instruments determined during Phases 3 and 4. This clearly illustrates that the phase model is not a linear one-off venture. The last three phases especially are repeated several times over in the course of a regional development strategy.

Evaluation

Regional development concepts are highly dynamic instruments. The evaluation of the development concept is therefore essential in attaining successful and sustainable development. Both the Project Hollandsche IJssel and the Ruhrtal Initiative update and confirm their action plans every year as a means of remaining flexible throughout the entire planning phase. Through this strategy, they can adapt to new

The regenerated Speke and Garston Coastal Reserve seen from above is, together with its sailing club and neighbouring business park, an important corner stone in the regional development of the Mersey Basin

situations as they arise, and integrate new possibilities and projects. By way of input for revising the action plan for the Hollandsche IJssel, the comparison of the development status against the plan is reverted back to the partner organisations and the steering group.

Funding

The means of funding dictates the speed and scope in which the project can be realised. In doing so it should be remembered that a good project idea is the best way to obtain financial assistance. Projects that also receive political support have an even better chance of receiving public funds. This having been said, projects often fail to obtain any public funding because they formally do not meet the administrative funding categories.

Regional co-operation is of increasing significance with regard to funding as the fund providers are demanding more and more co-ordination of regional needs. Through regional co-operation, individual projects and measures are incorporated into a broader overall context. As a result, they gain in significance and their 'critical mass' grows.

Alternative sources of funding should also be looked into at an early stage. To avoid any dependence on funds in short supply, it is worth looking around for partners who could develop into public-private partnerships. In doing so, it might well be possible to motivate those parties who would benefit the most from the local riverside being upgraded, to contribute funds or other forms of assistance.

When obtaining funding, it is vital that the potential partners share a common cause. A win-win situation should be created and the funding bodies and sponsors be shown the benefits of working in co-operation. In doing so, European funds can provide the necessary impetus to acquire additional public or private funds.

However, the regional development strategies and initiatives for sustainable regeneration of river landsca-

A historic river sail boat passing Loswall III: regenerated riversides make room for new business ideas

pes in the Artery partner regions are largely financed through public funds.

In the Ruhr Valley, the activities are supported through project-related funding from the state of North Rhine-Westphalia, the German Federal Government and through funds provided by the European regional development programmes INTERREG IIIB and Objective-2. In addition to this funding, aids also flow into development through public-private partnership agreements as well as local and regional administrations, authorities and associations. Collectively implemented regional projects are funded by the members of the Ruhrtal Initiative on the basis of a firmly agreed allocation formula. Projects at local level are supported by the relevant local authority or the project sponsor.

The Projectteam Hollandsche IJssel receives project-related funding from Holland's central government as well as sectoral funding for soil sanitation which is based on a soil sanitation programme scheduled to last several years. These funds are applied for and administered by the South Holland provincial government. When determining the size of funding, the anticipated profit resulting from the future use of the redeveloped riverbanks were taken into account as a way of keeping the use of public funding to a minimum. Furthermore, the riverside development projects are funded through private investments, assistance provided by local and regional authorities and funding from the European Union INTERREG IIIB programme.

The Mersey Basin Campaign, is supported by the British government by annually allocated funds. The allocation of these funds is based on a co-operation plan which comprises the annual targets of both the campaign and of the River Valley Initiatives for the individual river basins. On top of these allocations come funds from INTERREG IIIB, funding acquired through local partnerships and through direct and in kind support provided by the private sector.

The measures in the Stuttgart region are financed through public and private funds. In 2004, the Association Region Stuttgart has been enabled to support municipalities in their activities to implement the regional landscape park to the tune of 50 percent of the investment costs. The other 50 percent is covered by the relevant local authorities.

Success story:
Co-operative Funding The funding of the development strategy in the Ruhr Valley is shared among the partners. The allocation according to inhabitants and financial potential of the local authorities is as follows:

Allocation formula for Ruhr Valley funding

Bochum	25%
Hagen	20%
Hattingen	7.5%
Herdecke	3.75%
Wetter	3.75%
Witten	10%
Ennepe-Ruhr County	15%
RVR	15%

The Ruhrtal Initiative agreed on common funding for those projects which, given the nature of the partnership, were of significance or comprised areas involving several local authorities. To do so, a formula was compiled which was derived from and weighted on the basis of the number of inhabitants within a predefined area.

This is the model used to fund the agency 'The Ruhrtal'. In more precise terms, the better part of the required budget is made available through the state of North Rhine-Westphalia whilst the remaining 20 percent is allocated to the municipalities in line with the aforesaid legend. A similar situation applies to investments in the water trails which aim to develop new tourist attractions. In this context, the building and planning costs of the landing stages were collectively borne by the partnership.

This agreement binds the partners and bundles their initiatives. In turn, new possibilities for sustainably redeveloping the region result.

Regional Strategies

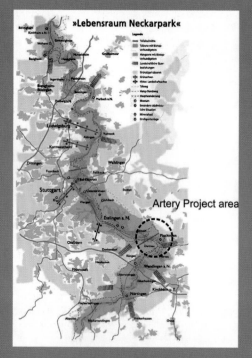

»Lebensraum Neckarpark«

Artery Project area

An overview of the Neckar-Landscape Park: the Regional Association Stuttgart (VRS) can now financially support municipalities in implementing this plan

Success story: Strengthening a Regional Authority The Verband Region Stuttgart (VRS, Association Region Stuttgart) is responsible for the regional planning and the Stuttgart Landscape Park in a region comprising 179 towns and municipalities. In order to lend this task its specific regional importance, the VRS has its own regional parliament which is elected directly by the local people.

The core element of the plan is a regional landscape park from which the Neckar-Landscape Park was derived. In the past, the VRS had virtually no means of implementing the Neckar-Landscape Park in spite of the political weight it carries. Therefore, it was dependent on the co-operative assistance of the local authorities as only they were able to assume the investment payments for the Neckar-Landscape Park. Many potential measures failed to materialise due to a lack of funding or a lack of landed property being made available.

Since early 2004, the VRS has received significant backing for its project as it can now exercise as implementing capacity for the Neckar-Landscape Park. This was achieved by virtue of amending the law which now enables the VRS to co-fund 50 percent of the measures required to implement the Neckar-Landscape Park. The other half must be provided through a local authority.

The implementation of the Artery pilot projects has been able to benefit from this new legislation.

Regional Development Strategy

~ Are the essential structural problems in the river basin being addressed?

~ In the process of doing so, has the core problem been highlighted and have serious solution strategies and measures been determined?

~ Are the envisaged objectives, visions and projects realistic?

~ Can the strategy truly improve the region's competitiveness?

~ Will every means and every resource in the river basin/region actually be used?

~ Do means of networking and forming alliances exist at regional, national or European level?

The projects in the Rhine-Neckar region have so far all been financed on the local level, be it through municipal funds, compensation payments for development schemes or co-financed through the federal state or Artery. With the establishment of the Association Region Rhine-Neckar and its responsibility for the implementation of the regional landscape park there now exists a possibility to directly allocate funds on a regional level.

As with Stuttgart-Neckar, no federal funds are available. At the same time, however, the regional authorities are not allowed to provide financial assistance for such investments. At present, the projects are half financed through the INTERREG IIIB funds and through the respective municipality where the project is taking place.

Text: Frank Bothmann

Review
and Prospects

Before regeneration the riverbanks in Ladenburg were hardly accessible and not very welcoming // An old, burnt out car – the area of
the Speke and Garston Coastal Reserve was
illegally used as rubbish dump

River Landscapes: Challenge and Potential

Neglected river landscapes and cities built with their backs to the river characterise many post-industrial regions in Europe. In these regions the populations are often unaware of the potential attractions of the urban river landscapes. Moreover, with the decline of the industries many job opportunities vanished and could not be replaced until today.

Long-lasting developments over the last century have deprived urban river landscapes of their former functions. Transport on rivers and water supply from rivers and industries on rivers, were once key factors for the emergence of cities, have now declined in importance. New dynamics developed in other places. What remained were polluted rivers and riverbanks. In some places they were even turned into dumping grounds, decommissioned industrial sites, inaccessible waterfronts, downgraded natural habitats. Eventually they became no-man's lands.

Since then regional development has made its contribution to a turn-around in many of the affected regions. European decrees like the Water Framework Directive and the European Spatial Development Perspective facilitated opportunities to attain a new quality in the regeneration of river landscapes. However, these new European standards still pose a challenge to various regions in Europe, namely the new Eastern European member states.

Riverside regeneration still holds a lot of opportunities for a region. River landscapes have wide-ranging potential which should not be left untapped for the development of attractive, economically vibrant and sustainable metropolitan areas – that is turning them into attractive tourist and recreation destinations, high-potential locations for new economic activities, multi-functional urban sites and valuable natural habitats. These possible developments can contribute jointly to the competitiveness of a city or a region.

The mobilisation of this potential requires a major conversion of river landscapes. Diverse measures are required. These include improving the accessibility to riverbanks, cleaning-up polluted soils and re-naturalising natural habitats. Recreation facilities for relaxation and for leisure sports like canoeing, biking, or hiking as well as new attractive business areas have to be created. Industrial and cultural heritage should be used as tourist attractions points.

Review

During more than three years of transnational co-operation (2003-2006), the Artery partners have implemented physical improvements in post-industrialised urban river landscapes on the rivers Ruhr and Neckar (Germany), the Hollandsche IJssel (Netherlands) and the river Mersey (United Kingdom). With financial support from the European INTERREG IIIB programme, North West Europe, the five regional leadpartners and their local sub-partners realised ten pilot projects, which all set a new benchmark for riverside regeneration by creating sustainable environments. Various measures, depending on specific regional circumstances, have activated latent potentials for urban-regional development by improving conditions for local people, for natural habitats and for

Additional investments facilitated through Artery (in millions of euros)

~ Mersey Basin Region	1.770
~ Hollandsche IJssel Region	10.190
~ Rhine-Neckar Region	0.200
~ Ruhr Valley	0.410
~ Stuttgart-Neckar Region	0.870
Total	13.440

local economies. Even more important than individual improvements is the fact that the projects triggered various complementary investments from other local participants, generating a set of coherent measures and providing a major boost for regional development. The Artery project has levered more than 13 million euros in extra funds, which is more than its original budget.

The leverage effect induced by the Artery project comprise in-kind investment contributions from private and public partners, extra funds or direct financial contributions to the pilot projects, the lease of land free of charge and also new investments in services or buildings directly supporting the Artery projects.

Finally, Artery has created new economic opportunities, new green spaces, and new leisure facilities, all geared towards encouraging local communities to 'turn around to face the water'.

The regenerated Speke and Garston Coastal Reserve is an ecologically valuable area. The local population can now use it for recreation

Additional jobs created or supported through Artery

~ Mersey Basin Region (supported)	9,000
~ Hollandsche IJssel Region (created)	31
~ Rhine-Neckar Region	0
~ Ruhr Valley (created)	28
~ Stuttgart-Neckar Region	0
Summe created	59
Summe secured	9,000

Key to success

The success for Artery was based on two principles. Firstly, project planners focused on broad stakeholder involvement. Secondly, the partners used advanced methods throughout their project development, benefiting from the transnational transfer of knowledge.

At an early stage, local populations were given the opportunity to participate in the planning process. Broad awareness-raising actions were directed at the general public and school children. Local businesses as well as major corporations were involved in public-private partnership co-operations. One major lesson was learned: many different players can contribute to the regeneration of derelict riversides. They can be municipal, private or other public landowners, local and regional development organisations, local residents, businesses benefiting from improved location qualities and creating new job opportunities, and nature conservation organisations. The involvement of these many players at an early stage when plans are still variable has proven indispensable.

A combination of local investments and transnational development methods contributed to this achievement and resulted in a mix of sophisticated participation measures. In the course of the Artery project, population groups have assumed responsibility for project implementation, businesses have contributed with money or in-kind for complementary investments, and various governmental organisations have provided unanticipated additional support. This ensured the sustainability of the project, giving these different players a sense of 'ownership' by sharing responsibility amongst them.

The different methodological approaches in the five regions and the three countries served to enrich the development of the ten projects. The four common themes acted as incubators for transnational method development: public participation, public awareness, public-private partnership, and regional strategies.

This would not have been achieved without INTERREG funding, nor without promotion, information, dissemination, discussion and persuasion, that is regional development management.

Future focus

Following the successful realisation of Artery from 2003 to 2006, the partners of Artery will continue their riverside regeneration measures in metropolitan areas, using approaches developed during the project.

The focus for the new activities will lie on the recreation, leisure and tourism markets. This offers new potentials for regional economic development. It will contribute to the quality of life of urban populations and re-connect them to their rivers.

Methodologically, the partners will build on strengths and experiences developed in the first phase, allowing selected pilot investments to become triggers for broader regional development, with contributions from other participants - population groups, private businesses, and other public bodies.

New Focus

The emphasis will shift to two themes:

~ The enhancement of sustainable regional economic innovation and competitiveness

~ The sustainable management and operation of converted areas and modernised facilities respectively

Sustainable economic innovation and job creation had already been touched on with Artery, but will now become the heart of projects, developing ways to implement the European Union's Lisbon and Gothenburg strategies at regional and local levels. This will improve competitiveness, sustainable development and the potential for innovation. In order for the individual improvements to be able to create these desired long-lasting effects, it is indispensable that they are kept well maintained and professionally operated. A major lesson learnt from Artery was the recognising importance of establishing sustainability for a project as early as possible. Means will be developed to ensure sustainable management of the new investments. The implementation of environmental regulations like the Water Framework Directive will also be a key issue.

Artery – Benefiting from Experience

The lessons learnt in Artery can be applied by other regional developers to their projects, whether it is a project for riverside regeneration or for regional development programmes. The methodological approaches from Artery help project planners to develop sustainable innovation.

Youth-Panel in Speke and Garston: Public Participation was an indispensable element of Artery // The new shallow water zone in Dossenheim near Heidelberg: Here children can now closely experience the Neckar

Review and Prospects

Other regions intending to develop their riversides will find the experiences of Artery to be a useful guide for the planning and development of their intended programmes. The Artery partners have profited widely from the interregional, transnational exchange. The authors are optimistic that all the know-how and the best practice generated during the programme can be found in this guide book. However, all partners are also open for a continuous discourse on the topic of riverside regeneration and active exchange with other European regions.

Artery II

The Artery partnership has benefited from growing mutual trust, maturing experience with European Union funds administration, constructive discussion and potential to learn from each other across borders. However, this relation-building is a process that takes time. Therefore, the true potential of these interregional relations has not yet been fully activated, the benefits not fully generated. Consequently, this partnership seeks a continuation of its co-operation through a new European project (Artery II) in the framework of the new 'Objective 3 – Territorial Cooperation' programme. This would also support the joint development of new approaches for the mentioned priority themes: the enhancement of sustainable regional economic innovation and competitiveness, and the sustainable management and operation of converted areas and modernised facilities.

A selective widening of the partnership shall increase the interregional knowledge transfer and will give other regions the opportunity to profit from the experience gathered in Artery I. Possible new partners could come from the new European Union member states or the current European Union accession candidates, or even other NWE regions.

Successful realisation of the project 'People to the River': The Artery partners celebrating in Mannheim

Artery Project Partners

INTERREG IIIB North West Europe

NWE Secrétariat
„Les Caryatides"
24, Boulevard Carnot
F-59800 Lille
www.nweurope.org

Region Ruhr Valley

*Regional Leadpartner/
Thematic Leadpartner*
'Regional Strategies'

Regionalverband Ruhr
Kronprinzenstraße 35
D-45128 Essen
www.rvr-online.de

Pilot Project Partners

Ennepe-Ruhr-Kreis
Hauptstraße 92
D-58322 Schwelm
www.en-kreis.de

Trägerverein
Unser Freibad am See Wetter (Ruhr) e.V.
Gustav-Vorsteher-Straße 36
D-58300 Wetter (Ruhr)
www.tv-freibad-wetter.de

Wittener Gesellschaft für Arbeit
und Beschäftigungsförderung mbH
Breitestraße 74
D-58452 Witten
www.wabe-witten.de

Rhine-Neckar Region

*Regional Leadpartner/
Thematic Leadpartner*
'Public Awareness'

Verband Region Rhein-Neckar
P7, 20–21 (Planken)
D-68161 Mannheim
www.vrrn.de

Pilot Project Partners

Nachbarschaftsverband
Heidelberg-Mannheim
Collinistraße 1
D-68161 Mannheim
www.nv-hd-ma.de

Stuttgart-Neckar Region

Regional Leadpartner

Verband Region Stuttgart
Kronenstraße 25
D-70174 Stuttgart
www.region-stuttgart.org

Region Hollandsche IJssel

Regional Leadpartner/
Thematic Leadpartner
'Public Participation'

Provincie Zuid-Holland
Postbus 90602
NL-2509 LP Den Haag
www.zuid-holland.nl

Pilot Project Partner

Gemeente Capelle aan den IJssel
Postbus 70
NL-2900 AB Capelle aan den IJssel
www.capelleaandenijssel.nl

Gemeente Krimpen aan den IJssel
Postbus 200
NL-2920 AE Krimpen aan den IJssel
www.krimpenaandenijssel.nl

Gemeente Nieuwerkerk aan den IJssel
Raadhuisplein 1
NL-2914 KM Nieuwerkerk aan den IJssel
www.nieuwerkerk-ijssel.nl

Ministerie van Verkeer en Waterstaat
Postbus 556
NL-3000 AN Rotterdam
www.verkeerenwaterstaat.nl

Stichting Molen Kortenoord
Postbus 193
NL-2910 AD Nieuwerkerk aan den IJssel
www.kortenoord.nl

Mersey Basin Region

Regional Leadpartner/
Thematic Leadpartner
'Public-Private Partnership'

Mersey Basin Campaign
Fourways House, 57 Hilton Street
UK-M1 2EJ, Manchester
www.merseybasin.org.uk

Pilot Project Partner

Liverpool Sailing Club
12, Buttermere Road, Court Hey
UK-L16 2NN, Liverpool
www.liverpoolsailingclub.org

Stockport Metropolitan Borough Council
Hygarth House, 103, Wellington Road
UK-SK1 3TT, Stockport
www.stockport.gov.uk

References

Alden J, Boland P (1996) Regional Development Strategies: A European Perspective. *Regional Studies Association, Routledge Taylor & Francis Group, London New York*

Allen W, Kilvington M, Horn C (2002) Using participatory and learning-based approaches for environmental management to help achieve constructive behaviour change. *Landcare Research Contract Report LC0102/057. Landcare Research, Lincoln*

Argote L, Ingram P (2000) Knowledge transfer. A Basis for Competitive Advantage in Firms. *Organizational Behavior and Human Decision Processes 82(1):150–169*

Arnaud JL (2001) Crossborder and Transnational Cooperation, the new Europe is inventing itself in its Margins. *Unioncamere & Notre Europe, Brussels*

Baker M (1998) Planning for the English regions: a review of the Secretary of States regional planning guidance. *Planning Practice and Research 13(2):153–169*

Bischoff A, Selle K, Sinning H (1996) Informieren, Beteiligen, Kooperieren. Kommunikation in Planungsprozessen; eine Übersicht zu Formen, Verfahren, Methoden und Techniken. *Dortmunder Vertrieb für Bau- und Planungsliteratur, Dortmund*

Coenen F (1995) Participation and the quality of environmental decision making. *Kluwer, Dordrecht*

Commission of the European Communities (2000) Directive 2000/60/EC of the European Parliament and of the Council of 23 October 2000 European Waterframework Directive. *European Communities, Brussels*

Commission of the European Communities (2002) Water Framework Directive. Annex I. Public Participation Techniques. *European Communities, Brussels*

Commission of the European Communities (2003) Directive 2003/35/EC of the European Parliament and of the Council of 26 May 2003 providing public participation in respect of the drawing up of certain plans and programmes relating to the environment and amending with regard to public participation and access to justice Council Directives 85/337/EEC and 96/61/EC. *European Communities, Brussels*

Commission of the European Communities (2004) Green Paper on Public-Private Partnerships and Community Law on Public Contracts and Concessions. *European Communities, Brussels*

Deutsches Institut für Urbanistik (ed) (2005) Public Private Partnership Projekte: eine aktuelle Bestandsaufnahme in Bund, Ländern und Kommunen. *Berlin*

Europäische Kommission (ed) (1999) EUREK Europäisches Raumentwicklungskonzept. Auf dem Weg zu einer räumlich ausgewogenen und nachhaltigen Entwicklung der Europäischen Union. *Potsdam*

European Union (ed) (2000) Water Framework Directive. *European Union, Brussels*

European Community Initiative INTERREG IIIB (2000) Spatial Vision for North-West Europe. *European Community, Brussels*

Gasteyer T (2003) Prozessleitfaden Public Private Partnership. *In: Bertelsmann Stiftung, Clifford Chance Pünder and Initiative D21 (eds) PPP für die Praxis. Clifford Chance Pünder, Frankfurt am Main*

Gramberger MR (2001) Citizens as Partners. OECD Handbook on Information, Consultation and Public Participation in Policy Making. *OECD, Paris*

De Haan G (2004) Politische Bildung für Nachhaltigkeit. *Das Parlament: Politik und Zeitgeschichte 7–8, Bonn*

Hachmann V (2004) Learning in Transnational Networks: the Case of the Community Initiative INTERREG IIC/IIIB. *Diplomarbeit Fachbereich Raumplanung an der Universität Dortmund*

Her Majesty's Stationery Office (ed) (2000) Public Private Partnerships: The Government's Approach. *Norwich*

Hildreth P, Kimble C (2004) Knowledge Networks. Innovation through Communities of Practice. *Hershey, London*

Kersting N (2004) Die Zukunft der lokalen Demokratie. Modernisierungs- und Reform-modelle. *Campus, Frankfurt*

Kollmann G, Leuthold M, Pfefferkorn W, Schrefel Ch (2003) Partizipation. Ein Reiseführer für Grenzüberschreitungen in Wissenschaft und Planung. *In: Kollmann G et al. (eds) Schriftenreihe Integrativer Tourismus & Entwicklung, vol 6. Profil-Verlag, München Wien*

McCloskey D (1994) 1760–1860: A Survey. *In: Floud R, McCloskey D (eds) The Economic History of Britain since 1700, vol 1. University Press, Cambridge*

Netherlands Ministry of Finance (2001) PPP Procurement Guide. *Ministry of Finance PPP Knowledge Centre, The Hague*

Nonaka I, Takeuchi H (1995) The knowledge-creating company. *University Press, New York Oxford*

Potter P (2004) Transnational Learning in Urban Regeneration. Methodologies of Knowledge Transfer in EU Thematic Networks. *University of Cambridge, Cambridge*

Sanoff H (2000) Community Participation Methods in Design and Planning. *Wiley, New York*

Schaffer F (ed) (1993) Innovative Regional-entwicklung. Von der Planungsphilosophie zur Umsetzung. *Lehrstuhl für Sozial- und Wirtschaftsgeographie, Universität Augsburg, Augsburg*

Schmidt E (et al.) (2002) Management Guide for Regional Co-operation. *Umweltbundesamt, Berlin*

The Regional Environmental Center for Central and Eastern Europe (2002) Developing Skills of NGOs, Public Education to Raise Environmental Awareness. *Szentendre*

UN Department of Economic and Social Affairs (ed) (1992) Agenda 21: Chapter 36 Promoting Education, Public Awareness and Training. *United Nations, New York*

United Nations Human Settlements Programme (2001) Focal Point Toolkit. Urban Governance Campaign. *UN-HABITAT, Nairobi*

Wolters C (2005) How to effectively design regional Development Strategies for Riverside Regeneration? *Documentation of the Artery workshop on November 17, 2005 in Essen. Regionalverband Ruhr, Essen*

Wolters C (et al.) (2005) Regional Development Strategies. Interim Report. *Regionalverband Ruhr, Essen*

Picture Credit

Unless otherwise indicated, references are made to sequential page numbers.

Page 4/5 left to right
Landesmuseum für Technik und Arbeit, Mannheim // Ruben Scheller, Nachbarschaftsverband Heidelberg-Mannheim (NV) // Die PR-BERATER GmbH, Agentur für Kommunikation, Köln //
WBW – Fortbildungsgesellschaft für Gewässerentwicklung // Nachbarschaftsverband Heidelberg-Mannheim (NV) // Nachbarschaftsverband Heidelberg-Mannheim (NV)

Page 7
European Commission, Directoriate-General for Regional Policy

Page 8, 9 right, **26** bottom, **29, 36** bottom,
48 top, **48** middle, **50, 62, 79, 90, 108, 110,**
111 left, **118** top, **129, 140, 156, 160**
Die PR-BERATER GmbH, Agentur für Kommunikation, Köln

Page 10, 151 bottom
Regionalverband Ruhr

Page 12, 20, 32 bottom, **55, 63** right, **65, 70, 76–78,**
80, 81, 85, 87, 94, 106, 111 right, **118** bottom, 151
middle, **164**
Mersey Basin Campaign

Page 16 top, **40** top, **60**
Guido Frebel

Page 16 bottom, **40** middle, **40** bottom, **42, 56, 74,**
123, 127, 131
Fotografie Ralf Breer, Hattingen; Rainer Scholz, Hagen

Page 15, 23 bottom, **166, 167**
Verband Region Rhein-Neckar

Page 23 top
Stadtarchiv Mannheim

Page 25
Gemeinde Altbach

Page 26 top
Avidrome Luchtfotografie, Lelystad

Page 30, 112
Mealeys Liverpool, Mersey Basin Campaign

Page 32 top, **120**
Stockport Borough Council

Page 34 top, **63** left, **66, 68, 69**
Peter Molkenboer

Page 34 middle
René Kandel

Page 34 bottom
Leendert Keij

Page 36 top, **36** middle
Gemeente Krimpen aan den IJssel

Page 38 top
Elly van Gelderen

Page 38 middle, **135, 141**
Henk van Veen

Page 38 bottom
Dick Groenendal

Page 44, 96, 104 top
Landesmuseum für Technik und Arbeit, Mannheim

Page 46 top, **113**
Foto Alex, Heidelberg

Page 46 middle, **46** bottom, **72, 92, 117**
Ruben Scheller, Nachbarschaftsverband Heidelberg-Mannheim

Page 48 bottom
Roland Appl

Page 51, 88, 126, 133, 143, 146
Frank Bothmann, Regionalverband Ruhr

Page 71, 84
Bureau Alle Hosper, landscape architecture and urban design

Page 82, 83
Johannes Kappler

Page 86
Wittener Gesellschaft für Arbeits- und Beschäftigungsförderung

Page 91 left, 93, 99–102, 104 bottom
WBW – Fortbildungsgesellschaft für Gewässerentwicklung

Page 91 right
Michael Johnston, Mersey Basin Campaign

Page 124, 136, 138
Peel Holdings Ltd.

Page 149
Gemeinde Edingen-Neckarhausen

Page 151 top
Projectteam Hollandsche IJssel

Page 153
Sarah Wallbank, Mersey Basin Campaign

Page 155
Colin McPhearson, Mersey Basin Campaign

Page 158
Verband Region Stuttgart

Page 161 left
LUZ Landschaftsarchitektur, Stuttgart

Page 161 right, 162, 163
Randall Thorp

Page 165
Institut für Umweltstudien (IUS), Heidelberg

Cover photo front
Peter Molkenboer

Cover photo back
Fotografie Breer; Scholz // Die PR-BERATER GmbH // Stockport Borough Council // Ruben Scheller, Nachbarschaftsverband Heidelberg-Mannheim // Foto Alex //WBW // Nachbarschaftsverband Heidelberg-Mannheim // Mersey Basin Campaign

In spite of intensive effort we were unfortunately unable to locate all photo copyright holders. Entitled claims will certainly be settled according to standard practice.

Index

Index

HOUSES AND GARDENS
OF CORNWALL

HOUSES AND GARDENS OF CORNWALL

A Personal Choice

by

Helen McCabe

With best wishes
Helen McCabe
June 26ᵗʰ 1990

TABB HOUSE

First published 1988

Tabb House, 7 Church Street, Padstow, Cornwall, PL28 8BG

Copyright © Helen McCabe 1988

ISBN 0 907018 58 0

Typeset and printed in Great Britain

for
Gilbert and Camilla

Acknowledgements

I SHOULD like to thank those owners who kindly showed me their homes and took the trouble to supply me with information about them. Their welcome and readiness to help increased the pleasure of writing this book. I am also grateful to many of them for the provision of photographs or permission to take photographs of their properties. I should like to include amongst their number Mr Michael Trinick, formerly director of the National Trust in Cornwall.

I am grateful to the committee of the Morrab Library, Penzance for allowing me to reproduce engravings, which are identified as illustrations numbered 1, 10, 18, 36, 38, 50, 64, 66, 73, 78, 84, 88, 89, and 105 in this volume, from books in its care; and to Mr Reginald Watkiss for photographing them so beautifully, besides providing me with illustrations, nos 99, 103, 104, 108, and 109. I am indebted also to the National Trust for its kind permission to reproduce nos 13, 14, 15, 16, 17, 19, 20, 21, 27, 55, 56, 59, 60, 61, and 72 and to Mrs Helen Jacoby for nos 23, 26, 37, 49, 65, 69, and 106. I am also grateful to Mr Michael Galsworthy for providing the photograph of Trewithen, reproduced on the front cover, and to Sir Arscott and Lady Molesworth St Aubyn for permission to use the painting of the four Misses St Aubyn by Arthur Nevis on the back cover.

I am indebted to Liz and John Bonython for introducing me to my publisher, Helen Jacoby for her practical help and hospitality during the course of my research and Carla Carlisle for her encouragement. Last but not least I should like to thank Caroline White and Ruth Lumley-Smith for their sound advice and editorial assistance.

Contents

Foreword

BY the eleventh century Cornwall was divided into districts known as Hundreds. The maps in John Norden's *Description of Cornwall* compiled in the early seventeenth century show nine divisions very clearly. *Stratton, Lesnewth, Trigg,* and *Pydar* run from the Tamar and Devon border down the north coast of Cornwall to *Penwith* in the extreme south west. *Kerrier, Powder, West,* and *East* stretch up from the Lizard and along the south coast to Plymouth. These distant-sounding Celtic names, their meanings shrouded in mystery, are my chapter headings.

I have included the principal castles of Cornwall in my survey, but it would have been impossible to describe every house and garden of note. Most of the important houses are mentioned but some are discussed more summarily than others. In the case of modest manor houses and farms, I have chosen both the unusual and the typical and those of which I am fond, and I have included one fishing village, as an example of a typical Cornish small coastal community.

Further details and histories of the families who lived in these houses may be found in the appendix.

There is also, for easy reference, a complete list of castles, houses and gardens mentioned in each chapter, stating if and when they are open to the public. Every castle and house is marked on the Ordnance Survey Map of Cornwall.

Complete List of Castles, Houses and Gardens

Those marked with an asterisk★ are NOT open to the public.

List of Illustrations

Key to map
Drawing based on John Norden's Topographical and Historical Survey of Cornwall 1610

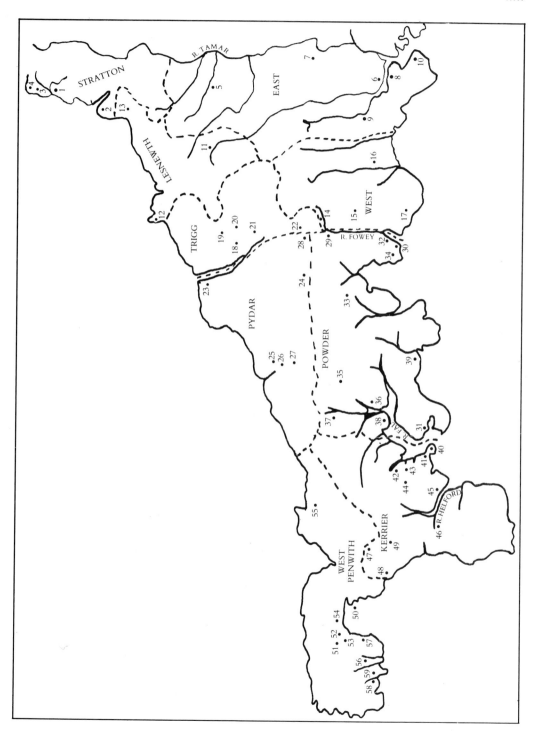

Introduction

CORNWALL is unlike any other British county, for the greater part of this long, south west peninsula is bounded by the Atlantic ocean. Its geography, archæology and climate have fashioned an independent breed of people with no great regard for their cousins across the Tamar. An understanding of Cornwall's Celtic past and dramatic history explains the modesty of some houses, the grandeur of others. The romantic inclinations of their owners are expressed in the gardens they designed and planted.

In Saxon times the Hundreds were sub-divided into manors under the jurisdiction of a local lord. William the Conqueror consolidated his conquest by strengthening this feudal system. All land ownership was vested in the king and many Saxon 'tons' became royal manors. William made his half-brother Robert, Count of Mortain, the 1st Earl of Cornwall. He gave the Earl the majority of manors in Cornwall which were held for him by Norman lords who then built castles to secure their position. Few of these French-speaking lords, many of whom held estates elsewhere, chose to reside permanently in primitive Cornwall where resources were meagre and the people spoke a language they did not understand.

In 1337 Edward III raised the earldom of Cornwall to a duchy and invested his seven-year-old son Prince Edward with the title of first Duke of Cornwall. Although the Black Prince occasionally entertained his lords and vassals and received his feudal dues, he never lived in any of his strongholds. Court influence did not exist in Cornwall. Throughout England, powerful noblemen such as the Earls of Warwick and Northumberland kept huge households, entertaining handsomely in the great halls of their castles and travelling with elaborate retinues to London in order to demonstrate their power and wealth. In Cornwall mediæval families simply cultivated their modest estates, the manors often little more than farmhouses.

With the Dissolution of the Monasteries in 1536, Cornish merchants and gentry were quick to acquire the monastic properties that came suddenly on the market. Those who supported the Reformation, enriched by the redistribution of monastic revenues and tithes, soon enlarged their old houses or built new ones. Port Eliot and Prideaux Place date from this time. A few families, the most notable being the Arundells of Lanherne, remained catholic and their estates soon fell into decay.

When relations with Catholic Spain deteriorated in the 1580s, Elizabeth I relied on the support of staunch Protestants such as Sir Richard Grenville and Sir Francis Godolphin. Although she approved unofficially of the Cornish gentry's verve in privateering and piracy against the Spaniards, she did not grace her distant and courageous subjects with a royal visit. There are no 'Prodigy' houses in Cornwall, fitted with great chambers and state apartments to receive a royal progress, but mainly unpretentious and homely Elizabethan manor houses, such as Trerice, tucked out of sight in sheltered coombes.

By the end of the sixteenth century political stability and increasing prosperity led to the 'Great Rebuild'. Manor houses were enlarged, with improved heating and lighting. The yeomen modestly followed the squires' example. This spate of new building gave considerable opportunities for local craftsmen to show their skills. The huge distances and poor roads forced Cornishmen to rely on local resources. Ash-grey granite from the moors of Bodmin and Penwith was the predominant building material, durable but extremely hard to carve and not suited to fine, decorative detailing. Elvan, the Cornish name for the local quartz-porphyry, was favoured by some Georgian builders for its finer texture, ideal for ashlar masonry. Antony and Trewithen are built of the light grey Pentewan elvan brought by sea and overland from quarries near Mevagissey. Hard Cornish slate, non-porous and quick-drying, made an excellent roofing material and was quarried early on in the Fowey and Padstow areas, the great Delabole quarry being worked as early as 1600.

Richard Carew, squire of Antony and a writer of note, stresses in his *Survey of Cornwall* of 1602 that the builder's aim was to construct solid, weather-tight buildings 'seeking therethrough only strength and warmness', for they had to withstand violent storms and shut out the moisture-laden air. Throughout history Cornish architects have fought an exasperating and uphill battle against damp walls and rotting roof timbers. The search for shelter from a furious and unpredictable wind dictated the sites of domestic buildings. Consequently, the fine houses built by the gentry are hidden in deep valleys or on the banks of an estuary rather than on the exposed sea coast. For practical more than æsthetic reasons the gentry planted trees where woodland did not already exist. Farms, too, were built on sheltered inland slopes rather than on coastal hill-tops where the old Celtic fortified castles had stood. Cottages were built of cob and thatch, for timber has always been scarce in Cornwall and good quality wood was required for boat-building. The rest was used for mine props and until the advent of coal for smelting fuel.

The gentry, Carew reports in 1602, 'keep liberal, but not costly builded or furnished houses, give kind entertainment to strangers, are reverenced and beloved of their neighbours, live void of actions amongst themselves. . .They converse familiarly together and often visit one another. A gentleman and his wife will ride to make merry with his next neighbour, and after a day or twain, those two couples will go to a third, in which progress they increase like snowballs, till through their burdensome weight they break again.'

Apart from relieving the monotony of country life in this remote part of England, visited by only the most intrepid outsider, one of the main objectives of the social round was match-making. As the history of the houses unfolds, one sees how closely connected by marriage are the Cornish gentry, a kind of respectable incest winding down all the family trees. But if match-making was a career for Cornish ladies, the men led busy, active lives as landowners, magistrates, military commanders, and members of Parliament. The gentry controlled local affairs.

During the Civil War Cornwall was the scene of heavy fighting. Most of the great Cornish families were ardent royalists and Charles I rewarded many of the men who had fought for him so bravely. Some were elevated to the peerage after the Restoration and enjoyed the King's favour. The sophistication of Charles II's court gave certain families a taste for splendour and refinement. Stowe, the magnificent house of Sir John Grenville, made Earl of Bath and Baron Grenville of Kilkhampton, was considered the noblest dwelling in the West of England.

No Cornish family was rich enough to build on the scale of Blenheim Palace or influential enough to entertain government members, as Sir Robert Walpole did so lavishly at Houghton Hall in Norfolk. The gentry were content to lead the lives of country squires, taking a keen interest in local politics and, more importantly, exploiting the mineral wealth found on their lands. In the latter part of the eighteenth century and well into the nineteenth, many old houses were pulled down and rebuilt with the profits from tin and copper. Fashionable architects from London such as Nash, Soane, and Wilkins were engaged to design these extravagant new houses in a variety of styles. For the first time, in a broad sense, ideas and fashions from the rest of England reached Cornwall, and with improved transport the Cornish themselves could travel further afield on excellent turnpike roads and in faster coaches.

The important houses of Cornwall are a monument to its history, for their owners helped to build it politically and economically. Modest houses, farms, and cottages complete the story. Architecturally provincial and behind the times they might have been, but some of these houses, rarely mentioned in books on English country houses, form a unique chapter in the history of English architecture.

GARDENS in Cornwall, like the houses, have a character of their own. Most are surrounded by woodland for protection against the wind, but the sea or river estuary can often be glimpsed through the trees. These trees may look poor specimens, stunted and misshapen, to a visitor from Wiltshire or Sussex, but they are cherished by their Cornish owners who have seen them do battle against the gales.

Spring comes early to Cornwall and the flowering season is long. Courageous snowdrops appear in January. By March daffodils cover fields and cliffs in a blanket of yellow warmth; gardens are bright with early camellias, raspberry to candy pink, and magnolias the size of water-lilies. Bluebells invade the May landscape, taking over hedgerows and meadows, a dramatic contrast to fire-red rhododendrons and azaleas. Hydrangeas bloom from August until November, their caps of blue, white, or pink lace compensating for lack of autumn colour: the wind curls and shrivels Cornish leaves before they can turn to burnished gold.

William Kent never came to this wild peninsula but his revolt against the 'artificial' in favour of a natural garden of winding paths, open vistas, and wooded glades inspired Cornish landowners to capitalise on the natural beauty of their native countryside. The sudden declivities, rocky chasms, and fallen tree trunks advocated by the Picturesque school are part of Cornwall's landscape. If the aim of the *jardin anglais* was to create an air of accident and surprise and to arouse varied sensations in the viewer, Cornwall's awesome cliffs, sublime seascapes, and exotic luxuriance contained all the drama and poetry needed to satisfy these demands.

Increasing world trade and travel brought to late eighteenth century Europe a flood of new and exciting plants. The emphasis was no longer upon design but on the creation of a flower garden. Humphry Repton, who visited north Cornwall, popularised the open terrace with surrounding flowerbeds overlooking the park, and paved the way for the plantsman's garden. Leading families in the county helped to finance the great plant collectors who journeyed to China and Asia at the end of the nineteenth and beginning of the twentieth centuries. The Cornish soil, fertile, quick-draining and rich in leaf mould, was ideally suited to nurture the seeds brought back from these expeditions. Magnolias, rhododendrons, camellias, and azaleas thrive in the damp, mild atmosphere and because frost is a rarity gardeners can grow in sheltered spots out-of-doors, sub-tropical plants that must be nurtured in greenhouses in other counties.

Today the great gardens of Cornwall are museums, storehouses of beauty where outstanding plants of every kind are catalogued, exhibited, and preserved. But Cornish gardeners are no mere custodians. Their success in creating new hybrids from the different species in their care, and their inventive attitude towards design make them innovators in the art and science of gardening.

Stratton

Call the hind from the plough, and the herd from the fold;
Bid the wassailer cease from his revel;
And ride for old Stowe when the banner's unfurled
For the cause of King Charles and Sir Bevil.
                    R. S. Hawker, 'The Gate Song of Stowe'

WHEN the Saxon King Egbert invaded Devon and Cornwall in 814 he entered a Celtic and largely Christian world that knew nothing of Roman ways. Saxon domination marked the end of Cornwall's long estrangement, both geographical and cultural, from the rest of England, and the birth of a new order. The Celtic form of clan ownership of land was replaced by the Saxon feudal system, whereby the county was divided into Hundreds. Saxon influence was first felt in the north of Cornwall, and in Stratton Hundred English place names mingle with the Cornish. Often an old Celtic settlement became a Saxon manor, with the Saxon word 'ton' or 'stow' added to the Cornish name, as in Kilkhampton and Morwenstow.

After the Norman Conquest, the estates were administered by tenants-in-chief in return for military support in time of war. Many manors were held by families who had come from Normandy, the Grenvilles among them. Although they have long since left Cornwall and virtually nothing survives of their seats, their presence remains strangely alive. The countryside along the north coast between Bude and Bideford, once their territory, has changed little. A number of steep and fertile valleys, cut into the high table-land at right angles to the sea, were lovingly described by Charles Kingsley in *Westward Ho!*

Each has its upright walls, inland of rich oak-wood, nearer the sea of dark-green furze, then of smooth turf, then of weird black cliffs which range out right and left far into the deep sea, in castles, spires, and wings of jagged ironstone. Each has its narrow strip of fertile meadow, its crystal trout stream winding across and across from one foot-hill to the other; its grey stone mill, with the water sparkling and humming round the dripping wheel . . . to landward, all richness, softness and peace; to seaward, a waste and howling wilderness of rock and roller, barren to the fisherman, and hopeless to the shipwrecked mariner.

Coombe is such a valley, and on the north slope, about a mile inland from the sea, the Grenvilles chose to build their home. **Stowe** was a modest, comfortable mediæval manor house for five centuries. In 1679 it was pulled down to make way for a new, grand residence worthy of an ennobled and powerful family. Now, Stowe Barton, a handsome late-eighteenth century farmhouse standing on the hillside in solitary splendour, is the only reminder of this great seat. The sloping fields bright with yellow furze in spring, the dense woodland, the squat tower of Kilkhampton church further up the valley, are as they were

Stowe

centuries ago. Only the huge, glaring-white telecommunication saucers tilted upwards and outwards to sky and sea on the moorland behind are out of place as we look back at Grenville history.

From the twelfth century onwards, the Grenvilles increased their power by adding to their lands and by participating in both local and national affairs. Like most Cornish gentry, they supported the House of Lancaster and were rewarded for their loyalty after the Battle of Bosworth in 1485, when Thomas Grenville became Esquire of the Body to Henry VII. Fifty years later, in 1536, Sir Richard Grenville acquired a great deal of land upon the Dissolution of the Monasteries. As steward of Bodmin Priory's extensive property in Devon, he was in an ideal position to purchase monastic tithes as well as monastic land, the value of which was to increase dramatically later in the century.

Roger Grenville, captain of the *Mary Rose*, was drowned when the ship capsized and his son, Sir Richard Grenville, also died for England, becoming a symbol of courage for generations to follow. ★

Sir Richard was succeeded by his grandson, Sir Bevil, who fought in the Civil War and died in 1645. His third son, John, inherited Stowe and continued to support his sovereign. ★★

Charles II rewarded John handsomely for his family's devotion to the Stuart cause. In 1661 he was made Earl of Bath, Viscount Lansdowne, and Baron Grenville of Kilkhampton and Bideford. He became Lord Lieutenant of Cornwall, Lord Warden of the Stannaries (with a pension of £3,000 a year paid out of tin revenues) and Steward of the Duchy, all lucrative appointments. It was small wonder that with such riches and a privileged position at Court - he was Groom of the Stole - he decided to pull down the old house of his forefathers and build a splendid new mansion in its place.

Although the lives of the various Grenvilles are well documented, very little written material or drawings of old Stowe have come to light, and Charles Kingsley's picture of a 'huge rambling building, half castle, half dwelling-house' with the 'lofty walls of the old ballium' still standing on three sides, the southern court having become a flower garden 'with quaint terraces, statues, knots of flowers, clipped yews and hollies', is probably almost entirely imagination. Whatever its charms, John Grenville evidently did not consider old Stowe either comfortable to live in or a suitable seat for an earl. In 1679 a grand new house built of red brick in the latest classical style was erected in its place.

Surprisingly little is known about new Stowe. Celia Fiennes, who journeyed through Cornwall on horseback in 1698, bypassed Kilkhampton on her way to Launceston and never saw Stowe, although she knew it had 'fine stables of horses and gardens'. A contemporary painting has survived, showing an impressive single block of four storeys with two projecting wings on either side. The roof is hipped, with dormer windows to light the top attic storey, while a grand staircase in the centre leads up to the ground floor. Its design resembles that of Clarendon House, Piccadilly, built for the great Lord Chancellor, Edward Hyde, by Roger Pratt in 1664, or Belton House in Lincolnshire, built twenty years later. Stowe was in the forefront of fashion and stylishness, massive yet serene with its straight elegant lines and symmetrical disposition. Its architect is not known but was undoubtedly a disciple of Inigo Jones and Roger Pratt. Such a sophisticated design – and on such a scale – together with the lavish use of building materials foreign to the county, were unknown in Cornwall at this time and suggest an outsider's hand. The magnificence of Stowe was unrivalled in the West of England.

Ironically, great Stowe's history is a short and sad one. John Grenville died in 1701, having enjoyed the house for twenty years, but henceforth personal tragedy brought about Stowe's demise. John's eldest son Charles shot himself accidentally while preparing to go to his father's funeral and was buried with him in Kilkhampton church. His small son Henry succeeded him as 3rd Earl of Bath but died of smallpox in 1711. A bitter family quarrel ensued, Jane and Grace Grenville, the 1st Earl's daughters, contesting the claim to the estates by their cousin George Granville. In the end, after much legal wrangling, George renounced his claim. Grace took over the Cornish estates and Jane those in Devon. Grace, created Countess Granville in her own right, briefly inhabited Stowe, but she married a Carteret whose seat was in Jersey. The mechanics of looking after her Cornish mansion became increasingly irksome. The expense was considerable and Stowe, for all its grandeur, was perishingly cold in winter. It must have been a hard decision to take, but in 1739 – just sixty years after its erection – she had it demolished.

Virtually nothing survives. The bricks and masonry, as well as fine

Stowe Barton (late 18th century) erected on the site of the stables at Stowe. The walls of the old carriage wash can be seen by the roadside.

panelling and plasterwork inside, were quickly taken and were incorporated into houses both in the area and further afield. The late seventeenth-century panelling and Grinling Gibbons-style carvings in the reading room at Prideaux Place, Padstow, together with a painting of *The Rape of Europa* by Antonio Verro above the chimney-piece, are traditionally held to have come from Stowe. The porch is now part of the town hall of South Molton in Devon.

Today a visitor looking at Stowe Barton, which has been erected on the site of the stables, may conjure up a vision of the great house with its proud cupola looking over the sea. The high walls of an oblong carriage-wash in front of the stables still stand and can be seen on the side of the road just below the farmhouse. Carriages gleaming in the sun can be imagined clattering through the entrance gate, past formally laid-out walled gardens with fountains and parterres, and up to the grand east front. Here visitors would alight, admiring the prospect over the valley towards Kilkhampton before ascending the elegant steps to the entrance hall. The coachmen would then have made their way round to the coach house by the west front.

THE Arundell family has left its mark in Stratton Hundred. **Ebbingford Manor**, outside Bude, was added to their estates in 1433, when Joan Durrant married an Arundell. For more than a century Ebbingford was one of the main Arundell residences. The 3rd Sir John Arundell died in 1561, probably at Ebbingford, and is buried with other Arundells in nearby Stratton church. His second son John set about rebuilding Trerice, the family seat near Newquay. He inherited considerable wealth from his father and married an heiress, Katherine

Coswarth, from the adjoining parish. After Katherine's death he married Gertrude Dennys of Holcombe, and it was possibly she who enlarged Ebbingford.

Originally Ebbingford consisted of a hall rising to the roof with, on the right-hand side, a bedchamber on the first floor and dining-room below. Separate buildings contained stables and kitchens. Gertrude added the stable block to the left of the hall, which was divided from the house by a slype or passageway. This block was subsequently converted into a chapel. The square, mullioned granite windows and huge chimney-stacks also date from Gertrude's time. After her death the newly-built Trerice presumably became the favourite Arundell residence, although Ebbingford was a useful outlet for their numerous children.

Richard Carew, who married Katherine Arundell's daughter Juliana, paints an attractive picture of Ebbingford in his *Survey of Cornwall*:

Master Arundell of Trerice possesseth a pleasant seated house and domains called Efford alias Ebbingford and that not unproperly because every low water there affordeth passage to the other shore; but now it may take a new name for his better plight for this gentleman hath, to his great charges builded a salt-water mill athwart this bay, whose causeway serveth as a very convenient bridge to save the wayfarer's former trouble, let, and danger.

The salt water mill can still be seen by the side of Nanny Moore's

Ebbingford Manor, Bude. The façade has 16th century mullioned windows (to the left of the porch) and Georgian sash windows (to the right).

Bridge in Bude. A large stone set in one corner bears the letters AJA 1589, for Anne and John Arundell.

In 1643 Ebbingford was occupied by Mary Arundell, the eldest daughter of the 5th Sir John. Her younger sister Anna had married Colonel John Trevanion, a troop commander in Sir Bevil Grenville's army. Trevanion and his soldiers camped at Ebbingford on the night of 15th May, 1643 before the battle of Stamford Hill near Stratton. Much merry-making eased the tension as tactics were discussed on the eve of the battle. Perhaps this was the last night that he and Anna spent together. Trevanion's victory at Stratton was short-lived, for he died, as Sir Bevil did, at Lansdowne that July.

After the Restoration very little is heard of Ebbingford. The newly created Baron Arundell of Trerice presumably let it to tenant farmers. When, in 1768, the 4th Baron died without issue, the Arundell estates passed to his wife's nephew, William Wentworth of Henbury in Dorset. A complicated settlement decreed that if William's son and daughter were to die without children the estates should go to the Aclands of Killerton, descendants of Margaret Acland, wife of the 2nd Baron of Arundell. This in fact happened and in 1802 Ebbingford, like Trerice, became the property of Sir Thomas Dyke-Acland. The Aclands never lived at Ebbingford, for part of it was already let to a Colonel Wrey I'ans. By this time Georgian sash windows had been put in and two brick chimney-stacks added. The other part of the house was a farm, and tradition has it that the first Methodist services in Bude were held in it in the 1820s.

In 1861 the Aclands gave Ebbingford to the Church Commissioners for use as a vicarage. It remained so until 1953, when the current vicar and the father of the present owner agreed to swop houses. Canon Walter Prest moved to 8 Falcon Terrace and Sir Dudley Stamp came to Ebbingford. It is truly remarkable that in 800 years Ebbingford has never been offered for sale on the open market. The walled garden shuts out the hurly-burly of Bude itself and embraces the house, as if to protect it from the ravages of time.

A FEW miles north of Stowe lies the village of Morwenstow, with its church, glebe farm, and vicarage. **Morwenstow Vicarage** was built in 1837 by Robert Stephen Hawker, the eccentric vicar of the parish who is remembered both for his poems and for his curious behaviour. Morwenstow is bleak and windswept. The cliffs nearby rise to almost four hundred feet and were notorious as a death trap for sailors. Hawker noticed that his sheep always used to shelter from storms on the ground immediately below the church, and so chose this site for his vicarage. A practical man who saw no need to employ an architect, he consulted T. F. Hunt's *Designs for Parsonage Houses* and adapted a model for 'a Clergyman's house on a moderate scale'. A solid, gabled

Morwenstow churchyard, just above the vicarage, with the white-painted figurehead of the wrecked Caledonia to the left.

Morwenstow Vicarage, built by R. S. Hawker in 1837.

Victorian gothic house, its chimneys are its most interesting feature.
They were all built to look like the church towers in parishes in which
Hawker had previously lived - Stratton, Whitstone, North Tamerton,
and Oxford. The kitchen chimney resembles his mother's tomb. Over
the front door is a verse inscribed in stone:

> A House, a Glebe, a Pound a Day,
> A Pleasant Place to Watch and Pray
> Be true to Church - Be kind to poor,
> O Minister, for ever more.

It is not a particularly attractive house, but some knowledge of
Hawker and his parish adds much to its interest. In 1823 he married
Charlotte I'ans, third daughter of Colonel Wrey I'ans of Ebbingford,
where no doubt the proposal took place. Although she was forty-one
and he was nineteen, they were devoted to one another and she
performed her duties as a vicar's wife extremely well. Hawker
described her face as 'a perfect image of noble Womanhood - oval -
blue-eyed - with a nose slightly curved somewhat like my own - a firm
mouth, and a forehead moderately high banded with soft light hair that
never turned grey to the last'. She played the piano, was fond of
literature, and above all, did not mind the isolation of this
'godforsaken' parish.

In 1864 Hawker wrote 'Mine is without exception the most desolate
parish for food in the country. We have no parish butcher - I remember
when there were three - no carrier or coach, not even a shop.'
Morwenstow folk had to grow their own produce and buy the
occasional extras from travelling pedlars. Hawker's eight hundred or
so parishioners were mainly farm labourers, uneducated and pitifully
poor. 'How they exist is a mystery,' comments Hawker, 'especially
after the potato blight.' The agricultural wage was eight shillings a
week, hardly enough for rent, fuel, and clothing, after paying for their
frugal diet of bread, potatoes, garlic, and tea.

Hawker was a compassionate man and his letters are full of details
about the appalling housing conditions in his parish. Most families
lived in one room, which was damp, dirty, and devoid of sanitation.
Open cesspools were the norm. Typhus and cholera were common and
child mortality high. The villagers searched the beaches for wreckage
in the hope of finding something that would help to relieve their dismal
existence.*

Hawker was prone to melancholia and however much he busied
himself with parish affairs felt miserable and lonely after Charlotte's
death. In 1864 he married Pauline Kuczynski, a half-English, half-
Polish woman of twenty, forty years his junior. He met her when a
Yorkshire vicar brought her family to Morwenstow to convalesce.
Pauline came with them as nursery governess to the children. She and

Marsland Manor. The gatehouse.

Marsland Manor. The mid-17th century court.

Hawker were devoted to one another and she bore him three daughters before he died in 1875.

MORWENSTOW is hardly a village in the accepted sense, since there are no houses round the church save for the vicarage and farm. A mile to the north, however, just south of Marsland stream which runs into the Atlantic and marks the Cornwall-Devon boundary, is the **manor of Marsland**. Built of the local rubblestone in the late sixteenth or early seventeenth century, it forms four sides of a square. A hall, parlour, cellar, front and back kitchens, and a salting house (for the salting down of meat) are grouped round the court. On the gable above the salting house are the initials W.A. and the date 1656 carved on a stone. This part of the house was rebuilt in the mid-seventeenth century when modernisation was deemed necessary. A fine new bedchamber was made in the upper storey, the date 1662 being visible in the plaster.

A small gatehouse forms one side of the court, a reminder that the concept of fortification had not yet disappeared in this remote corner of England. A narrow entrance way would keep intruders out.

In Kingsley's *Westward Ho!*, Henry Davils of Marsland taught Amyas Leigh 'how to catch trout when he was staying down at Stowe'; another friend, Will Cary, galloped off to Marsland Mouth to round up two Jesuit priests for Sir Richard Grenville. Perhaps this tiny manor was their home.

2.

East

This triple-crowned mounte, though abandoned,
retayneth the forme, but not the fortune and favour of
former times.

John Norden, *Description of Cornwall* (1610)

THE year after William I's victory over King Harold at Hastings, the Norman army was sent to quell a rising in Exeter. After a brief siege, the Normans marched further west, across the river Tamar, which forms the eastern boundary of East Hundred and of Cornwall itself. The 1st Earl of Cornwall chose Launceston (or Dunheved as it is called in Domesday Book) as the administrative centre of his domain. The simple earth and timber castle he built on the summit of the hill gave him a superb vantage point. No hostile force could approach from Bodmin moor to the west or from Dartmoor to the east without being detected.

Although by the end of the twelfth century military architects felt that the fortification of a larger area by means of a massive curtain wall with several entrances provided a more flexible type of defence, giving room to launch a counter-attack, **Launceston Castle** did not follow this trend. Surprisingly, given the strategic importance of the site, no further building took place until well into the thirteenth century when Richard, younger brother of Henry III, became Earl of Cornwall in 1227.

Richard modernised the castle, which he made his headquarters, and built most of the stonework we see today. He constructed a shell keep on the motte or mound, replacing the earlier wooden palisade with a circular stone wall. Lean-to buildings were constructed round it on the outside, to provide basic accommodation on the mound, while behind the wall, within the keep itself, the soldier now had a safe platform from which to fight. At a later date a round tower rising one storey higher was built inside the shell keep, the narrow passage between the two being roofed over at wall-walk level. This increased the width of the wall-walk and meant the ground floor premises were better protected. The joist holes for this roof can be seen half-way up the tower. The wall-walk survives in part, although the parapet has almost gone, and one can see in the thickness of the wall the remains of the two staircases which led up to it. On the west side of the keep is a large recess which contained one or two *garderobes* or privies which were ingeniously flushed out in wet weather by the drain from the wall-walk.

The tower, built of a dark-coloured shale, is entered by a pointed arch doorway on the west side. The ground-floor room has no windows and was presumably used for storage. A staircase leads up to the first-floor room which is well lit by a pointed window. There is also a fireplace with the remains of a slightly projecting hood corbelled out

Launceston Castle (mid-13th century). The shell keep and gatehouse.

from the wall.

Access to the shell keep was strongly defended. A roofed-over staircase with high walls on either side and a gatehouse at the foot led up the precipitous south side of the motte. To reach this bottom gatehouse an intruder would first have had to cross the moat that protected the south and east sides of the castle, and break through the massive south gate, now the main entrance to the castle from the town. Within this gate lies the bailey of the castle, known as Castle Green ever since it was turned into a landscaped park in Victorian times. The west side of the bailey was protected by a steep slope and curtain wall, part of which still survives to the west of the south gate.

The Black Prince's survey of the castle, made in 1337 and the earliest surviving documentary evidence of the stone structures, makes it quite clear that the buildings within the bailey were the domestic and administrative quarters of the castle. There were three separate halls. The largest was the assize hall, where petitions were presented and feudal dues paid. Here, wrong doers were punished and flung into the nearby gaols. This hall was still used 'for syses and sessions' when John Leland, Henry VIII's librarian, visited Launceston in 1539, and was better maintained than the other buildings because of its legal function.

The second hall, with walls 'of timber and plank', together with a chamber and small chapel, formed a suite of rooms for the use of the Earl. Because the Earls of Cornwall so rarely came to Launceston, these fell quickly into disrepair, and had disappeared by Leland's time. Earl Richard is supposed to have spent Christmas here, shortly before his coronation as Holy Roman Emperor at Aachen in 1257; and to have raised troops in Launceston for the battle of Lewes in 1264. After his death in 1272, the Earl's quarters lay empty for a long time. The third

Launceston Castle. The South Gate.

and smallest hall, 'convenient for the Constable', with a chamber and cellar, survived until the seventeenth century.

Two prisons are mentioned in the 1337 survey, one of which may be the small dank room opening off the north gate, later known as the Doomsdale Tower. The Quaker, George Fox, was imprisoned here in 1656 for refusing to take off his hat to the judge and pay the fines for the illegal distribution of religious tracts. He and his two companions also refused to pay their gaoler the outrageous sum of seven shillings a week for food and were kept in the filthy Doomsdale Tower for five months for daring to protest against such extortion. A modern plaque commemorating their bravery was placed above the doorway in the north gate by the Society of Friends. Fox was not the only prisoner to have suffered for his faith. Many Catholic recusants including Francis Tregian were imprisoned at Launceston in the latter part of the sixteenth century for their adherence to the Church of Rome.

The remains of the kitchens, ovens, pantry, and brewhouse can be seen to the west of Castle Green. The deer park which once supplied venison for the Earl's table, as well as timber for repair work, has disappeared. By the mid fourteenth century the whole castle complex was in dire need of repair. Pigs had trampled down the moat and weakened the stone foundations, but neglect and lack of funds to carry out the necessary renovations were the chief cause of decay.

Launceston held out for the King during the Civil War, finally surrendering to Fairfax's army on February 25th, 1646, but by then the only habitable part of the castle was the gatehouse, possibly the south gate, containing two rooms in which the constable lived. The castle never again performed a military role. It was so near to collapse that the Parliamentarians did not bother to demolish it, as they did Pontefract Castle in Yorkshire, in accordance with their general policy. When the assizes and county gaol were moved to Bodmin in the 1840s, Launceston Castle ceased to be the centre of Cornwall's legal administration.

THE Normans built another fortress near Saltash. They had to be able to repel raiders coming up the Tamar from Plymouth Sound, and **Trematon Castle**, standing two hundred feet above Forder creek on the north bank of the Lynher, commands extensive views up and down stream. Domesday Book records that a Norman lord, Reginald de Valletort, held Trematon Castle, together with thirty-three manors forming the 'honour' of Trematon, for Robert of Mortain. The Valletort family remained in control of these lands until 1270 when Richard, Earl of Cornwall, consolidated his position by adding Trematon to his other Cornish strongholds.

Trematon Castle. View from the river Lynher, from Lysons' Magna Britannica.

The site of Trematon Castle - the name comes from the Cornish *tre* (homestead, town) and *matern* (king's) - may well have been known to the Romans and Saxons. Pieces of pre-Norman carved stone have been found within the castle precincts, suggesting the existence of an early encampment. The building of the present castle was begun by Robert of Mortain in 1080 and completed in the mid thirteenth century. It is of

Trematon Castle. The late 13th-century gatehouse.

the same 'motte and bailey' design as Launceston. The motte, an artificially raised mound built on existing rock, supports the oval shell keep built by Richard, Earl of Cornwall. Its walls are thirty feet high and ten feet thick. Nothing remains of the buildings inside the keep, but holes for the beams and stone supports for the roof timbers or vaultings bear witness to the Earl's private quarters which consisted of a hall, lodging chamber, and kitchen.

A large part of the protective curtain wall stretching out from the keep still stands. This once surrounded the entire bailey and formed a triangular-shaped courtyard of almost an acre. Stables, quarters for the Earl's retinue, kitchens, a hall and a chapel were built within this enclosure. A fine, square gatehouse built by Edmund, Earl of Cornwall, towards the end of the thirteenth century has also survived. At ground level there are dungeons and rooms with slits in the walls for bowmen, as well as two portcullises. The accommodation was up-to-date and luxurious. Each room has a stone fireplace comprising two slim columns with foliated capitals on either side of a hooded lintel. The mason's desire to make the fireplace decorative as well as functional suggests the presence of a royal visitor. The Black Prince occasionally came to Trematon and would have lodged here rather than in the primitive quarters inside the keep.

In 1549 Trematon Castle was attacked by Humphry Arundell, leader of those Cornishmen who objected to the first Book of Common Prayer introduced in Edward VI's Act of Uniformity. This forbade the use of Latin and the saying of Mass. Richard Grenville, a staunch supporter of the Crown, was forced to withdraw to Trematon, ready to withstand a siege. His lily-livered men, however, clambered over the castle walls at dead of night and made their escape. Alone and unarmed, Grenville met the rebels at the small sally gate which still stands. One honourable soldier was no match for Arundell's unscrupulous troops and, as Carew reported some fifty years later, 'those rake-hels stepped between him and home, laid hold on his aged and unweyldie body and threatened to leave it lifeless if the enclosed did not wave their resistance'. The seventy-year-old veteran was ignominiously carted off to Launceston gaol while Trematon was ransacked. Women 'were stripped from their apparel to their very smocks and some of their fingers broken to pluck away their rings'.

In 1580 Sir Francis Drake is said to have stored a huge amount of silver at Trematon. This treasure had been seized from Spanish ships during Drake's sea voyages and needed a secure resting-place until it was either shipped or trundled in wagons to the Tower of London. By

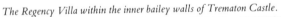

The Regency Villa within the inner bailey walls of Trematon Castle.

now, the castle had, generally speaking, outlived its usefulness. Twenty years later, Carew noted in his *Survey of Cornwall* that 'all the inner buildings falleth daily into ruin and decay only there remains the ivy tappised walls and a good dwelling for the keeper and his gaol'. Neither Roundheads nor Cavaliers made use of Trematon during the Civil War. It was beyond repair.

The ultimate fate of the castle precincts is an unusual and happy one. In 1807 Benjamin Tucker, a naval man who was appointed Surveyor General to the Duchy of Cornwall, secured a ninety-year lease on the castle from the Prince Regent, part of the agreement being that he would spend £1,000 on the building of a house within the inner bailey walls. It is not known whom Tucker chose as his architect, but whoever it was proved sensitive to the romantic qualities of the site. He built a simple, rectangular Regency villa whose elegant, straight lines and smooth stuccoed walls - now a delicate soft beige colour - are the perfect foil to the round, roughly-hewn stone walls of the castle keep which towers to one side of it. Windows along one side of the house look out at the old castellated curtain wall leading up to the keep. Today a well-stocked herbaceous border runs its length and a lawn slopes down to the gravel sweep in front of the porch. The other façade faces Edmund's gatehouse, its verticality an effective contrast to the long, low proportions of the villa. Tucker pulled down a section of curtain wall on either side of the gatehouse because it blocked out the potentially superb view. Now, in the distance, you can see the wooded slopes of Antony on the other side of the Lynher and watch ships move slowly past.

In 1846 Queen Victoria was cruising with Prince Albert in the Royal Yacht *Victoria and Albert*. They took a boat trip up the Tamar in the yacht's tender *Fairy*, and scores of onlookers lined the river bank in the hope of catching a glimpse of their sovereign. When she saw Trematon high up on its mound she noted with satisfaction 'that it all belonged to Bertie as Duke of Cornwall'. Trematon still belongs to the Duchy, and like Tucker before them, the present owners are lessees. They are enthusiastic tenants who have put their hearts into the upkeep of both house and garden. The latter is full of surprises. An orchard of apple and pear trees leads into an Italian garden with romantic-looking sculptures placed incongruously in the grass or next to a crumbling mediæval wall. The steepness of the paths, either up to the keep or down to the picturesque creek and village of Forder, reminds us of the impregnability of Trematon and its former military role; the gravel walks and flowerbeds are proof of Trematon's domestic *raison d'être* today.

The river Tamar, a natural boundary between Devon and Cornwall, was not merely important from a military standpoint. Before the advent of the railway it provided the main mode of transport for everything this part of Cornwall needed. The roads were much used by trains of pack animals and were frequently impassable in wet weather. Sheet maps of Cornwall were neither accurate nor detailed, and since sign-posts were scarce the chance of taking a wrong turn in the tangle of high-hedged lanes was high. The river, which twists through the countryside like a benevolent snake, was a more reliable highway.

The numerous quays along its banks were served by sailing barges which were able to tack up and down river. Coal and limestone were brought up river from Plymouth to feed the lime-kilns on these quays, lime being needed to fertilise the arable land upon which the economy of every estate depended. The agricultural produce of the farms and market gardens was sent down river cheaply and efficiently. The Tamar also offered the Cornishman relatively easy access, by boat to Plymouth, to the rest of England. Journeys by river and sea were quicker and safer than those made by road.

Quite apart from the practicality of living close to the river, the wooded valleys provided shelter from cold winds as well as picturesque views across the water. Some of the loveliest and largest houses in Cornwall are built on the banks of the Tamar.

Cotehele, the most important Tudor house in the county that now belongs to the National Trust, stands high above the river near Calstock. The maze of lanes that criss-cross the rolling fields do not lead to a grand approach. You stumble upon Cotehele by chance. Its granite walls are hidden from sight among trees, its semi-fortified appearance bidding the stranger keep his distance.

The house takes its name from Hilaria de Cotehele who married a Devonian, William Edgcumbe, in 1353. William and his descendants lived at Cotehele for the next two hundred years, by which time the Edgcumbes had become one of the leading families in Cornwall. Hilaria's house was a small, neat square. The rough sandstone rubble walls and round-headed lancets on either side of the gatehouse in the middle of the south wing probably date from her time. It was not until the end of the fourteenth century that extensive alterations were made.

After an eventful life the 1st Sir Richard Edgcumbe retired to Cotehele where he lived until his death in 1489. He built the buttressed barn and present granite gatehouse in the conventional Gothic manner, with a pointed arch entrance and castellated parapet round the top. He was also responsible for the upper floor round the inner Hall Court and for a new chapel with a bellcote and finials of moulded granite in a corner of the Retainers' Court. Here, the two-light pointed windows beneath square dripcourses also date from Sir Richard's time. The imposing pointed archway into the Hall Court, with a porter's squint in the wall on the left of it, is thought to have been the main entrance in Hilaria's day, since traces of a road leading to it have been found beneath the meadows. ★

Sir Richard redesigned the approach from the south side. Under the gatehouse a narrow cobbled passage, wide enough for a pack horse, leads into the Hall Court. This court was completed by Sir Richard's son Piers, who had married an heiress called Joan Durnford. She owned a considerable amount of land on both sides of the Tamar, so Piers Edgcumbe could easily afford to complete his father's proposed scheme for the enlargement of Cotehele. Building continued until his death in 1539, the great hall which one enters from the hall court being his major contribution.

One of the present delights of Cotehele is that the original house and contents have been preserved virtually intact. In 1553 Sir Piers's son Richard built a house in the park at Mount Edgcumbe on Plymouth Sound, part of his mother's Rame Head property. This became the Edgcumbe family's main seat. Henceforth Cotehele was only occupied spasmodically, the ideal retreat for widows or elderly relatives. No further additions were made except for the north-west tower in 1627. This was still castellated in the Gothic tradition, but its windows are square-headed. Apart from the east front, which was reconstituted for a widowed Countess of Edgcumbe in 1862, the rest of Cotehele,

Cotehele. The Retainers Court, showing the late 15th century chapel built by Sir Richard Edgcumbe and the entrance into the Hall Court, with a porter's squint to the left.

including the furnishings, has remained undisturbed. Much of the furniture is exactly as it was when stiff-ruffed ladies and gentlemen walked through the rooms chattering, laughing or complaining of the cold.

The design of Sir Piers Edgcumbe's great hall, which contrary to mediæval tradition has no screens passage, illustrates how behind-the-times Cornish builders often were. Hammerbeam roofs had evolved in the fourteenth century, the most famous example being that of Westminster Hall (1397-9), and although most commonly found in East Anglia, became popular elsewhere. Henry VIII favoured a hammerbeam roof when he built the great hall at Hampton Court in the 1530s. But at Cotehele, which was built only a decade earlier, the roof is of a type fashionable in the fourteenth century, the great hall of Penshurst Place in Kent (1341-8) being similar in appearance. The slope of the roof is divided by three horizontal beams or purlins which are strengthened by moulded arched supports. These windbraces interlock to form a uniform pattern which rises in four tiers. The absence of any tie-beams adds to the feeling of height. A neighbour, Henry Trecarrel, used the same technique for the hall of his manor house, Trecarrel, now the barn of a farm a few miles south of Launceston.

The surprisingly small kitchen at Cotehele is contemporary with the hall and full of interesting equipment. The flour ground in the estate mill (which can be visited) was stored in a covered wooden hutch;

Cotehele. The Hall.

butter and salted meats were kept in pottery crocks, while a huge pestle and mortar was used for pounding meat and herbs, as well as for breaking up blocks of salt and sugar. Pot-hooks with adjustable hangers for the various cooking vessels hang in the huge hearth. On a wintry day the fires in the hall and kitchen are lit and it is easy to imagine the cook keeping an eye on the heavy pots over the hearth or putting her feet up on the settle when her work was done. The ovens in the middle of the wall were used for baking bread, cakes, and pies. A fire of dry sticks or furze would be lit and the red hot embers raked out when the oven was sufficiently heated. The dough would then be pushed into the oven with a long wooden peel or flat shovel, and the door sealed with clay.

Cotehele is particularly rich in textiles. The plastered walls of most of the rooms are hung from cornice to dado with seventeenth or early eighteenth century Flemish and English tapestries. These colourful alternatives to wood panelling helped to shut out the draughts. They were cut to fit the walls and often adjacent panels did not match in terms of subject matter. In the White Bedroom, for example, there are three Mortlake tapestries depicting mythological scenes, two Flemish panels showing a crowing cockerel and figures in armour, and another English fragment illustrating a maritime scene.

All the needlework at Cotehele is in a remarkable state of preservation because wear and tear from the seventeenth century onwards was minimal. The walnut four-poster bed in the White Bedroom is hung with typical Jacobean crewel embroidery, soft blue, musky pink and green stylised flowers and trees in wool being mounted on linen. The crewelwork curtains, similar to those on the famous Abigail Pett bed in the Victoria and Albert Museum, depict a lively hunting scene with deer and hounds on the lower border. The bed hangings in the South Room - the solar of the Tudor house with a squint looking down into the hall - are of red wool worked in stem stitch, back stitch and coral knot mounted on linen. Lilies and oak leaves form the basis of the pattern. There are also superb examples of seventeenth century *gros point* embroidery as well as silk hangings and canopies.

If time seems to have stopped inside the house, the garden shows more recent activity. Much of the planting was done in late Victorian times. In the north-west corner of the house is the meadow, approached through an archway in the Retainers' Court. Nineteenth century prints show cows being milked here; now it is planted with a mass of daffodils. Nearby is the upper garden, surrounded by a mediæval wall covered with jasmine and forsythia. Fuchsias and peonies, regal purple and shocking pink, flank the path beside a water-lily pond and on the lawn below are a magnificent tulip tree and golden ash.

A path leads past the Jacobean tower and round the side of the house

Cotehele. The Punch Room.

to the east front. You pass a group of white thorn and whitebeam trees, an old Judas tree, a mulberry, a cork oak, and a silver weeping lime, as well as fiery rhododendrons and milky white wistaria. On the three terraces that descend in steps below the east front purple polyanthus flower in spring, scarlet roses in summer. Two huge magnolias spread their branches over more steps leading down to the valley garden. It is worthwhile pausing here to admire the view. Crimson rhododendrons and deep pink camellias overhang a thatched Victorian summerhouse in front of a mediæval pond; beyond is the old dovecote.

The lower part of the valley, full of birdsong, is planted with spindly spruce and larches but also contain rarer specimens such as *Davidia involucrata* (known as the Dove Tree because of its large white bracts which hang in pairs), and *Gunnera manicata* which resembles a gigantic rhubarb plant. All these trees and shrubs protect the garden from east winds.

Communications were so bad in Tudor times that the estate had to be more or less self-sufficient. The buildings clustered round the quay, which was abandoned at the turn of the century, are a unique reminder of industrial activity at Cotehele. The National Trust has dug out and repaired the little docks and restored the warehouses, one of which has been turned into a small museum explaining the river economy. Half a mile up a stream that runs into the Tamar just below the quay is the manor mill, where corn was ground to provide flour for the house. This has been restored and is once again in working order, like the horse-powered cider mill beside it. Outbuildings nearby contain the blacksmith's forge and the wheelwright's and saddler's shops. You can even see the sawpit where wood was sawn into planks ready for the carpenter.

COTEHELE is such a haven of peace and beauty that it is hard to appreciate why the Edgcumbes decided to live at **Mount Edgcumbe**. They could not have chosen a more different site. Whereas Cotehele is protected by woods and is invisible to the outside world, Mount Edgcumbe is as vulnerable as it could be, standing open and exposed on a hill overlooking Plymouth Sound. Perhaps the Edgcumbes were tired of being hemmed in, and relished the thought of unlimited space, magnificent sea views, a larger, more modern house and, if Cotehele was too inaccessible in bad weather, a nearby town.

Piers Edgcumbe's son, Richard, came into a substantial inheritance on the death of his father in 1539, and eight years later began to build a mansion in the enclosed deer park. A local mason called Roger Palmer from North Buckland built him an oblong house which was still mediæval in appearance since the front was not fully symmetrical. It had round, battlemented towers at each corner and a magnificent hall lit from the top by clerestory windows. This huge room invited the admiration of all who saw it. Richard's grandson, Richard Carew, writing in 1600, describes how this hall 'rising in the midst above the rest, yieldeth a stately sound as you enter the same. In summer, the opened casements admit refreshing coolness, in winter, the two closed doors exclude all offensive coldness.' The height and light of such a room probably thrilled the Edgcumbes after the dark, low-ceilinged rooms at Cotehele.

From the parlour and dining-room they could look across the Sound towards Plymouth or watch the boats sailing up the Tamar towards Saltash. Rich meadows and fertile arable land 'abundantly answereth a housekeeper's necessities', Carew goes on, while 'a little below the house, in the summer evenings, sein boats come in and draw with their nets for fish, whither the gentry of the house walking down, take the pleasure of the sight, and sometimes, at all adventures, buy the profit of the draughts'. The admirals of the Spanish and Dutch fleets whom Sir Richard entertained in Queen Mary's reign must have told their countrymen of the splendours of Mount Edgcumbe, for forty years later the Spanish Duke of Medina Sidonia was apparently determined to live there once the Armada had conquered Britain.

In the early nineteenth century Mount Edgcumbe was a well-known beauty spot which visitors from all over England came to admire.

Portrait of Richard Carew, squire of Anthony House and author of The Survey of Cornwall *1602.*

Tragically, those very qualities that made Mount Edgcumbe such an agreeable dwelling place – its hilltop position and proximity to a sea port – were the cause of its destruction. It was gutted by a German incendiary bomb in April 1941, during a particularly heavy raid on Plymouth, and a large part of the contents, including a number of portraits by Reynolds which hung in the dining-room, went up in flames.

A rebuilding programme was launched in 1958, but compensation was not sufficient to allow a reconstruction of the eighteenth century west wing. The Mount Edgcumbe of today resembles the original, smaller sixteenth century edifice. The sturdy outer walls and octagonal towers (which replaced the original round towers in the eighteenth century) managed to withstand the bombardment and were quickly repaired but the interior had to be entirely rebuilt. The hall is still centrally placed and lit from above, but an elegant main staircase now leads from it to a gallery at first-floor level. The cornices, doorheads, and chimney-pieces of the principal reception rooms on the ground floor are of such superb craftsmanship that it is easy to believe they are late Georgian in the company of the fine eighteenth-century furniture, which survived the onslaught.

By 1789, when George Edgcumbe became the first Earl of Mount Edgcumbe, the formal lower garden by the water's edge and the long avenue from the ferry up to the house had all but disappeared, George's brother, the previous Lord Edgcumbe, having abandoned formality for a more naturalistic, picturesque lay-out. In the 1750s he built a ruin overlooking the sea with a tower and an authentic-looking gothic window, not to mention plenty of ivy, moss, and lichen on its walls. Zig-zag walks were laid out nearby and a vast park was planted with the trees that we see today. George Edgcumbe and his son, Lord Valletort, continued the romanticising process and were responsible for the final plan of the park and gardens, which have scarcely altered since.

Uvedale Price, whose *Essay on the Picturesque* (1794) radically affected nineteenth-century English garden design, thought Mount Edgcumbe was a 'wonderful place'. Detesting the monotonous 'clumping and belting' of trees in Capability Brown's artificially-created landscaped parks, Price praised the natural grandeur of Mount Edgcumbe with its spectacular setting, impressive views and sudden contrasts. Seats and buildings including follies were positioned throughout the park not only to introduce different associations to the mind but to allow the visitor to pause and enjoy each successive vista. Fanny Burney experienced the delight and excitement of these sudden contrasts, when she visited Mount Edgcumbe in 1789 as maid of honour to Queen Caroline. 'In one moment you might suppose yourself cast on a desert island, and the next find yourself in the most fertile and luxuriant country'.

A guide-book published in the 1820s entitled *A Walk Around Mount Edgcumbe* takes the visitor on a tour of the grounds. It is still possible to identify most of the attractions described. The tour begins in the upper park on the high ground to the south of the house and follows the main drive as it twists through the woods and along the coast. The first section was known as the Great Terrace, high above the sea, the next Picklecombe, 'a little valley so regularly scooped out by nature as almost to bear the appearance of art'. Further on, an evergreen plantation led to a stone arch 'like a perforation of the natural rock' and into a number of walks snaking up and down the hillside. Plenty of benches were provided for the visitor to sit and marvel at the views. The trees are now thicker and taller here so that some of the vistas have vanished, and the zig-zag walks have almost gone. The drive continues through more woods to emerge into an open park with the gothic ruin as one of its focal points.

If the park and woods, with their dark dells and hidden glades, provoked gloomy, awesome thoughts, the adjacent pleasure gardens brought nothing but delight. An Ionic rotunda on the edge of a wood overlooking the water was a favourite place by which to dally. This is the Temple of Milton, whose bust adorns a niche inside. On the wall is an inscription from *Paradise Lost*:

> Overhead upgrow
> Insuperable height of loftiest shade
> Cedar and fir and pine and branching palm
> A sylvan scene and as the ranks ascend
> Shade above shade, a woody theatre
> Of stateliest view.

It is as if Milton had Mount Edgcumbe in mind when he wrote these lines.

Beyond, on the lower slopes of the park leading down to the sea, lie three formal flower gardens. The rambling English garden is predictable enough, with fine specimen trees and shrubs dotted about the lawns and traditional herbaceous borders. The French garden, in contrast, 'is a little square enclosure bounded by a high cut hedge of evergreen oak and bay and laid out in a parterre, with a basin and *jet d'eau* in the midst, issuing from rock-work intermixed with shells'. The parterre consists of impeccably clipped low box hedges arranged in a pattern and crammed with flowers - bright yellow and orange wallflowers in spring, red begonias in summer.

Nearby is a small rock garden full of surprises. Fragments of columns and capitals from Greece and Rome stand cheek by jowl with tiny tombstones in memory of animals beloved by the Edgcumbes. Should the visitor wish to rest, he can pause at Thomson's Seat, a pedimented alcove with Tuscan columns dedicated to the eighteenth-century poet James Thomson. Thence to the Italian garden: 'This plot of ground,' says the guide-book, 'is disposed in a regular manner with gravel walks, all meeting in the centre at a basin of water, in the midst of which is a beautiful marble fountain.' At one end of the garden two diagonal flights of steps lead to a white balustraded terrace decorated with classical ornaments. The graceful silhouettes of Apollo Belvedere, the Medici Venus, and Bacchus stand out against the dark green hedge of ilex and bay behind. An eighteenth-century orangery flanks the opposite end of this cool and elegant retreat.

Mount Edgcumbe illustrates the divergent trends of eighteenth-century garden design. It is part Garden of Artifice embellished with 'Gothick', 'Classick' and 'Rustick' buildings and planted with decorative flowering shrubs from far and wide and part romantic garden planted with native trees, and designed to appear untouched by human hand. These 'wild' places together with the craggy coastline that forms a grand but natural boundary to the gardens, made Mount Edgcumbe a rival to Stowe and Bleinheim as a tourist attraction.

Today the **Country Park**, as it is called, belongs to the Plymouth City and Cornwall County Councils and is open all the year round for

Antony House 1711-21. The south front and forecourt. Engraving from Natural History of Cornwall (1758) by William Borlase.

local citizens to enjoy. Trees and shrubs have been planted on a vast scale - a thousand camellias in 1976 - to mark a new stage in the development of these historic gardens.

ACROSS the water, to the north-west of Mount Edgcumbe, lies **Antony House**, the home of the Carew family since the late fifteenth century. In the 1540s, at about the same time as Sir Richard Edgcumbe was building his great house at Mount Edgcumbe, Thomas Carew married Sir Richard's daughter Elizabeth. Their son Richard married Juliana, daughter of Sir John Arundell of Trerice.

A portrait of Richard Carew, painted when he was thirty-two, hangs in the hall at Antony. Unlike many of his Elizabethan contemporaries, who chose to sit for their portraits in sumptuous apparel, Richard is dressed in black and wears a gold chain around his neck. His hair is close-cropped and his eyes, at once sensitive and alert, are those of an intelligent man. Characteristically, he holds a book in his hand. (See p. 28)★

Carew was a loving father - he and Juliana had ten children - and nothing seems to have given him more pleasure than to watch them grow up. He paints a charming, albeit idealised, picture of life at Antony. One of his favourite pastimes was fishing. He describes how he built a salt water pond on the shore of the Lynher to which he would retreat whenever possible. The pond was surrounded by a palisade to keep out scavenging otters, and had a flood-gate in the corner nearest the sea to let in the salt water.★★

Although the remains of Richard Carew's 'fishful pond' can still be seen, the house he knew has vanished. Richard's great-great-grandson William Carew pulled down the old Antony and built a fine early

Antony House. The north front.

Georgian mansion in its place. Building began in 1711 and was completed in 1721. Tantalisingly, the name of the architect of this supremely elegant house has never been discovered. For a long time Antony was attributed to James Gibbs, architect of several London churches and some fine country houses such as Ditchley in Oxfordshire, but no mention is made of Antony in the documentation of Gibbs's work, nor is there any record of his having worked in Cornwall. However, a design in Gibbs's *Book of Architecture*, published in 1728, closely resembles the plan and elevation of Antony, so it seems reasonable to suppose that a pupil or follower of Gibbs may have been the architect.

The design of Antony is immensely satisfying. It consists of a two-storey solid block whose proportions and size are perfect, neither too long nor too high, large yet not too large. This simple shape, which gives the house its gravity and serenity, is softened in a number of ways. Both the south and north fronts have a central pediment which projects very slightly, stressing three out of the nine bays of each façade. There are two rows of straight-headed windows on each façade and no external ornament at all. The punctuation is subtle. The space between the three centrally placed windows is narrower than that between the group of three on either side. The eye is thus made to focus on the centre, while the simplicity of the overall design allows us visually to absorb the whole. The six dormer windows in the hipped roof are not placed exactly above the bays below but slightly to one side, introducing small out-of-key lines into the otherwise vertical and horizontal grid.

If both façades follow the same architectural pattern, they are completely different in feeling. The south or entrance front forms one side of a square forecourt. At right angles to it on each side are two low wings of red brick. These wings are connected to the house by a brick wall with quaint corner pavilions. A wrought-iron entrance gate in the middle of the fourth side is aligned with the main entrance of the house, now camouflaged by a porte-cochère added after 1838. The contrast between the ornamental red brick and the soft, pale grey of the unadorned façade, built of Pentewan stone from Mevagissey, lessens the formality of the approach and introduces a note of gaiety. The friendly little courtyard welcomes and protects, while the north or garden front commands. Unfettered by buildings on either side, it stands majestically on a terrace that leads down to a magnificent stretch of lawn. If you turn back to look at the house, you look *up* at it, as if to pay homage.

In the last decade of the eighteenth century the fashionable landscape gardener Humphry Repton made one of his famous Red Books for Antony. Due to improved roads and modes of transport, Cornwall was no longer the backwater it had been. Cornish gentry were as up-to-date and receptive to new trends as their counterparts elsewhere. They could now be leaders of fashion instead of followers. The Earl of St Germans had consulted Repton in 1792-3 and commissioned him to lay out the grounds of Port Eliot, a few miles further up the Lynher river. Repton realised that there were other clients for him in this hitherto uncharted territory, and lost no time in persuading the Carews to consult him. He was immensely excited by the picturesque possibilities afforded by the riverside sites of both properties.

Repton's Red Books always contained one sketch showing the property in its current state and another showing what it would look like if his designs were implemented. The lodge at Antony Passage and the lodge at the main gate resemble his drawings, but many of his ideas for embellishing the terrace and approaches to the house were not taken up. His basic landscaping principles were, however, followed (whether then or later is not known), and the result is a tribute to him. The impressive expanse of grass which stretches down to the water's edge, broken by pleasingly positioned groups of trees, creates a grand yet natural vista of great beauty. Shelter from the elements is so essential in Cornwall that few houses are allowed the privilege of a magnificent prospect. The view from the terrace at Antony is one of the most memorable.

The interior of Antony, as harmonious as its exterior, has altered little since the house was built. Most of the rooms on the ground floor are panelled with Dutch oak, whose rich brown colour gives them warmth and intimacy. The main staircase, also of Dutch oak, has

Antony House. The hall and staircase.

Portrait of Charles I by Edward Bower.

turned balusters and Corinthian columns for newel posts. These are lit at each turn by their original bubble lights, early eighteenth-century blown glass globes.

With the exception of one or two Elizabethan pieces in the hall, which probably came from old Antony, most of the furniture is contemporary and some of it has never left the house. The flavour of each room is unmistakably English – sober, restrained and sensible. The Carews were enthusiastic and discerning patrons of English portraitists and craftsmen. On the staircase hang portraits of the Carew, Pole, and Coventry families (Lady Anne Coventry married Sir William Carew, the builder of the house) by Kneller, Beach, Northcote, Lely, and Hudson. Two full-length portraits of the 8th and 9th Earls of Westmorland by Reynolds hang in the saloon. But the most moving picture is that of Charles I by Edward Bower, thought to have been a pupil of Van Dyck and employed to sketch the King during the trial proceedings at Westminster Hall. Bower shows him dressed in black and clutching his silver-topped cane. He looks dignified and authoritative, yet his tightly pursed lips and inquiring eyes betray his apprehension.

A FEW miles upstream, on the west bank of the river Tiddy which flows into the Lynher just south of the village of St Germans, stands **Port Eliot**, the seat of the Earls of St Germans. The name is thought to come from St Germanus of Auxerre who is known to have visited Britain in the fifth century, although there is no firm evidence that he was in Cornwall. At all events, after the Saxon subjugation of Cornwall, King Edward the Elder established a Cornish diocese here with St Germans as its cathedral. In 1170 the priory of Secular Canons was replaced by a strict Augustinian community, for whom the existing church, with its grand Romanesque portal of seven concentric rings set beneath a pointed gable, was built.

The earliest reference to the old priory as *Porte Ellyot* occurs in 1573, ten years after John Eliot, a merchant adventurer engaged in trade in Plymouth, bought the land and priory buildings from the Champernowne family for £500. John Champernowne had been quick to seize the main chance when, to quote Carew, 'the golden shower of the dissolved abbey lands rained well near into every gaper's mouth'. After the required bowing and scraping, Henry VIII leased the house and demesnes of St Germans to Champernowne in 1540.

His son, Henry, made over the property to John Eliot.

The quaint almshouses which flank the main street of St Germans were built in John Eliot's time. It was an age when merchants wanted to affirm their status and their reputations as public benefactors, as well as to erect a personal memorial by building almshouses. Eliot was evidently anxious to be remembered by posterity, for he stipulated in his will that the fine carved panelling he had installed in some of the

Elizabethan almshouses (restored) in the main street of St Germans.

rooms of *Porte Ellyot* was not to be removed.

John's nephew Richard inherited, but it was his successor, another John Eliot, who increased the family's wealth by marrying a rich yeoman's daughter with sizeable lands. ★

Until the end of the eighteenth century, both church and house stood above marshes that bordered a tidal creek. However Edward Eliot, created 1st Lord Eliot in 1784, decided to reclaim an area of park from the mud flats and by the time he consulted Repton in 1792 the operation was well under way. He had already planted extensively on the ridge of high ground overlooking the river to the north and in the pleasure grounds on the promontory to the east. When discussing his plans for Port Eliot in his Red Book, Repton acknowledged Lord Eliot's skill and 'the tuition of that Judgement, Taste and persevering Energy which have not only clothed the naked hills with flourishing plantations, removed mountains of earth and vast beds of rock . . . but the waters of the neighbouring Ocean, converting into a cheerful Lawn that which was occasionally a bed of ooze. . .'

Today mature beeches and flowering shrubs adorn the slopes of the promontory, and the grand sweep of parkland culminating in the highpoint of the ridge known as the Craggs is as impressive as Repton would have wished. But sensibilities have changed. Whereas Repton's 'powers were subdued' on viewing the 'sublime horrors of the Craggs', we enjoy the prospect for its gentle natural beauty.

Repton's Cornish clients were stubborn individuals, for here, as at Antony, many of his recommendations were rejected. He wanted to link the house and church, in accordance with his principle of creating a harmonious whole, and proposed 'a cloister gateway' consisting of a billiard room above and a passage leading to the family pew below. Lord Eliot dismissed this idea and ten years later, in 1802, called in the

Port Eliot. 19th century engraving of the west front, designed by Henry Harrison c. 1829.

London architect John Soane to remodel the house. Soane had already carried out alterations to Lord Eliot's London house in Downing Street, and Eliot was sufficiently pleased with the result to summon him to Cornwall, initially to advise on repairs to the church which had partially collapsed.

Lord Eliot died shortly afterwards in 1804 and it was left to his son, created Earl of St Germans in 1815, to supervise operations. Soane entirely remodelled the east front, which had originally consisted of the prior's hall and parlour. The house had already been modernised in the Georgian era. Soane retained the sash windows and mansard roof but gave Port Eliot a mediæval air, in keeping with its origins, by adding a battlemented parapet all the way round. At this time the east side of the house was the entrance front, the approach being from the quay and through the pleasure grounds. As time went on access by land became more convenient, and the 2nd Earl employed the architect Henry Harrison to make the present day west-entrance front in the 1820s.

A DESCRIPTION of **Trerithick**, a small manor house half way between the hamlets of Altarnun and Polyphant just west of Launceston, will give a more balanced view of East Hundred, where so many of the great Cornish families chose to live. It was built by a well-to-do farmer

Trerithick, Altarnun. The entrance porch bears the date 1585

called John Hec in 1585, as the inscription above the entrance porch records. The ground plan follows the traditional mediæval pattern of hall, parlour, cross passage, and service room, but the rooms are lighter and more spacious than usual. Plain, flat-headed windows have replaced the round-headed type common in Tudor times, a few of which can still be seen in the outbuildings scattered round the house. The entrance porch, no longer a pointed Gothic arch but flatter and more rounded, has heavy roll-moulding running up the jambs and around the spandrels. The local stone carver has placed a single ball inside each spandrel. Granite lends itself to these bold, uncomplicated motifs; their honest simplicity gives this Cornish farmhouse its charm.

Houses like Trerithick, of limited value architecturally, are most revealing historically. They are the product of the Great Rebuild, that period in English history c. 1580-1620, when a rise in population and an increase in wealth resulted in a building boom. When there was cash to spare, the middle classes chose to add on and improve, rather than to build new houses. Here, the two-storeyed hall was abandoned in favour of a single-storey room with a bedchamber above; large fireplaces replaced the open hearth; more domestic rooms were added at the back of the house and larger windows, glazed rather than shuttered, were installed as time went on. Half veiled by the high stone hedge of an impossibly narrow lane, solid and unpretentious, Trerithick is an integral part of the Cornish heritage.

3.

Lesnewth

and on the night
When Uther in Tintagil past away
Moaning and wailing for an heir, the two
Left the still King, and passing forth to breathe,
Then from the castle gateway by the chasm
Descending thro' the dismal night – a night
In which the bounds of heaven and earth were lost –
Beheld, so high upon the dreary deeps
It seem'd in heaven, a ship, the shape thereof
A dragon wing'd, and all from stem to stern
Bright with a shining people on the decks,
And gone as soon as seen. And then the two
Dropt to the cove, and watch'd the great sea fall,
Wave after wave, each mightier than the last,
Till last, a ninth one, gathering half the deep
And full of voices, slowly rose and plunged
Roaring, and all the wave was in a flame:
And down the wave and in the flame was borne
A naked babe, and rode to Merlin's feet,
Who stoopt and caught the babe, and cried 'The
King! Here is an heir for Uther!'

Alfred Lord Tennyson, *Idylls of the King* (1859)

THE savage, jagged cliffs and wild countryside round Tintagel have excited both the admiration and the awe of travellers over the centuries. John Norden, in his *Description of Cornwall* written in the first decade of the seventeenth century, finds the village of Tintagel 'a poore decayde place, furnished with a few decayde howses', and Botreaux (now Boscastle) where once stood a fine castle, 'a meane market towne and enhabited for the moste parte by poore men'. **Tintagel Castle**, once impregnable, is now 'rent and ragged by force of time and tempestes'. Indeed most buildings in Lesnewth, according to Norden, have fallen into decline and poverty with the demise of these two mighty seats.

More than two hundred years later, a nineteenth-century observer, Dr Maton, finds the surrounding countryside just as bleak and rugged but 'the whole forms such a dismal picture of desolation that we began to imagine ourselves removed by enchantment out of the region of civilisation'. If the dark, cavernous inlets and lashing waves depressed and terrified poor Norden, they thrilled the romantic mind which yearned to lose itself in the immensity of untamed nature.

Tintagel, with its ancient ruined castle half on a cliff, half on a rocky headland, is a supremely romantic site. The restless sea forever pounding against the castle walls reminds man of his helplessness in the face of the elements; the brutal grandeur of the scene raises it from the ordinary to the sublime. Fear was a favourite romantic emotion, and the very real danger of a shipwreck lured many a nineteenth-century

The ruins of Tintagel Castle (mid-13th century).

tourist to Tintagel's treacherous shore. Today it is the myths surrounding Tintagel Castle as well as the dramatic setting which draw the crowds.

The actual history of Tintagel is as fascinating as the legends which have grown up around it. The headland provided good pasture land for the earliest settlers, who farmed there until about 350 A.D.. But when St Juliot, a Celtic missionary thought to have come from South Wales, arrived in the sixth century, this farming community had already gone. Excavations of the monastery he founded have revealed the remains of a sophisticated complex of buildings which included an infirmary and libraries. There is evidence that the settlement had running water and a primitive form of central heating. A Celtic granite tomb shrine which houses relics or possessions associated with St Juliot was uncovered in one corner of a Norman chapel built over the monastic site in the twelfth century. Coins, such as a silver penny of the reign of King Alfred the Great (871-99 A.D.), and fragments of 'red-gloss' pottery imported from Gaul or made in Britain, bear witness to a simple but flourishing monastic community whose life revolved round labour in the fields, services in the chapel, and the teaching of the peasants. Saxon domination brought peace and prosperity to an end, the monks left and the headland was once more uninhabited.

The Normans recognised the importance of the site. Building was begun by Reginald de Dunstanville, one of Henry I's illegitimate sons and Earl of Cornwall, in the 1140s. The chapel on the highest point of the headland, almost 100 metres above sea-level, and the walls of the great hall date from this time. Geoffrey of Monmouth, writing in the twelfth century, states: 'Tintagel Castle is situated upon a massive headland surrounded by the sea on every side, with but one drawbridge entrance to it, through a straight rock which three men shall be able to defend against the whole power of the kingdom'.

On Reginald's death in 1175, the castle was leased to local lords. No further building took place until 1236, when Richard, Earl of Cornwall and brother of Henry II, repossessed the castle and enlarged it. He built two wards on the mainland, the upper one containing the round keep, and constructed a massive curtain wall which embraced both island and mainland. The mainland wards provided accommodation for the earl's soldiers as well as stabling for the horses; the island quarters consisted of the great hall, eighty feet long, the private apartments of the earl, and the kitchens and buttery. By mediæval standards it offered an unbeatable defensive position and comfortable living quarters.

Like other Norman strongholds in Cornwall, Tintagel was in a sorry state when the Black Prince, as 1st Duke of Cornwall, claimed it in 1351. Part of the mainland site was used as a state prison in the late fourteenth century, but after that the buildings began to crumble. John Leland, who travelled all round Cornwall, reported in 1535 that 'shepe now fede within the dungeon grounde and rabbits abounde' but that the drawbridge was still there. Soon afterwards, part of the slate cliffs must have collapsed and sent the wooden drawbridge crashing into the sea. When Norden arrived on the scene in about 1600 he had to clamber down a perilously steep path to cross onto the island and up an equally precipitous cliff on the other side.

The leaste slipp of the foote sends the whole bodye into the devouringe sea and the worste of all is the higheste of all, neare the gate of the entrance into the Hall where the offensive stones so exposed hange over the head, as while a man respecting his footinge he endangers the head and looking to save the head endangers the footing.

Even sheep, he comments morosely, who climb the cliffs like goats, sometimes miss their footing and hurl headlong into the sea.

Elizabeth I, anxious to strengthen coastal fortifications against the possibility of a Spanish invasion, ordered Sir Richard Grenville to survey Tintagel Castle, but although he reported that the ruins could be made fit for occupation the work was never carried out. Decay continued until the mid nineteenth century when an energetic vicar, the Revd R. B. Kinsman, took the parish of Tintagel in hand and cleared away the undergrowth which by then had covered the ruins.

Tintagel has been associated with King Arthur since the twelfth century, when Geoffrey of Monmouth, in his *History of the British Kings*, set King Arthur and his court, in the castle. In reality, King Arthur was probably a sixth-century Celtic chieftain who bravely rallied his countrymen to resist the Saxon invaders, but such was the popularity of both Monmouth's story and Sir Thomas Malory's *Le Morte D'Arthur* (which drew heavily on Monmouth), published in 1485, that most people came to think of Arthur as a mediæval king whose gallant knights embodied the ideals of chivalry and honour. He has remained a hero in English eyes, a symbol of goodness and

A romantic, 19th-century view of Tintagel Castle.

strength. Tennyson revived the legends in verse in his best-selling *Idylls of the King* (1859), and kept Tintagel as the traditional birthplace of Arthur.★

The early Norman church of St Merteriana still stands on the bleak cliff, buffeted by the wind, and from the doorway you can look down at the castle on the headland and out to sea. To a rational mind it is merely a majestic view; for a lover of poetry, music and legend, the ruins of Tintagel Castle inspire tumultuous thoughts and dreams.

THE main street of Tintagel, mostly a depressing mixture of late Victorian and Edwardian dourness and modern garishness, brings us back to reality with a bump. But in the same street the **Old Post Office**, a small but remarkably preserved mid fourteenth-century manor house reminds us of our Plantagenet past. Both walls and roof are built of local slate, now mellowed to an attractive grey-brown. The square chimney-stcks, the main one built in three tiers, are capped with four up-ended slates, apparently the local pattern of chimney-pot but now rarely seen. The roof bulges and dips because the heavy slates have made the supporting beams sag.

Inside, a passage runs from the porch to the tiny garden at the back of the house. A door to the right leads into the parlour, possibly the original kitchen, with a narrow staircase leading to the bedchamber above. The occupants would have enjoyed gazing up at the wooden rafters bent into a comforting arc above their heads. An even narrower stair made of slate slabs set into the wall leads to a gallery overlooking the hall on the other side of the passage. This small area, lit by a tiny window and cut off from the hall by a wall of timber and plaster, was probably another bedchamber. To the left of the entrance passage is the

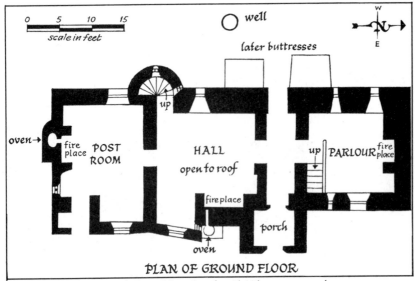

The Old Post Office, Tintagel. Ground floor plan of a mid-14th century manor house.

hall, two storeys high and open to the roof timbers. The hearth has a huge grey-blue slate overmantel, an attractive contrast to the white-washed walls. A spiral staircase leads to a third bedroom, below which is the Post Room, complete with counter and receiver, just as it was in Victorian times.

In the early nineteenth century there was no post office in remote Tintagel. Letters were delivered on foot from Camelford, some six miles away. When Sir Rowland Hill introduced his Penny Postage in 1840, more and more people wrote letters more and more frequently – to the inconvenience of the overburdened postman trekking across the hills from Camelford. So in 1844 the GPO set up a Letter Receiving Office in Tintagel, and rented one room for the purpose from the owner of this old manor house. It served as the village post office for the next fifty years, until the then owner decided to sell the house. The Old Post Office, as it was then called, was put up for auction in 1895. An artist, Miss Catherine Johns, bought the house, as she was alarmed at the thought that it might be pulled down and replaced by a tawdry hotel. This was becoming the fate of many other cottages, for Tintagel was already a tourist Mecca for Victorian romantics acquainted with the poems of Tennyson, Arnold, and Swinburne. Sales of pictures by a group of fellow artists were held to raise money for repairs. The National Trust purchased the Old Post Office in 1903 for £100 and has cared for it ever since.

HIGH on a nearby hill, above the coast road from Tintagel to Boscastle, stands the even older stone farmhouse of **Trewitten**, dating from the thirteenth century. It is a late version of a long house, having two storeys, 'two up and two down'. The larger ground-floor room

The 14th-century church house, Poundstock.

provided living quarters for the family, with the later addition of a fireplace and stone newel stair in a rectangular turret to reach the room above, while the smaller downstairs room was originally a cattle byre.

FURTHER up the coast, a mediæval **Church House** survives in Poundstock, a picturesque village a mile or so from the sea. Built in the late fourteenth century, it is a two-storeyed rectangular building complete with stone buttresses (added early this century) and four-light casement windows made of oak. In order to raise money for the little church tucked away in the trees above the church house, the church wardens used to give revels in the upper room. Beer was brewed downstairs and sports were held in the graveyard behind, there being no gravestones at that time. Mummers' plays were part of the festivities and fairs were also held, the merchants setting up their stalls in the room downstairs and round the walls. In the sixteenth century the church house became the parish almshouse, and today the upper floor is still used for village meetings and to celebrate Harvest Festival. For some reasons church houses are peculiar to this area of Cornwall only, and none save this one in Poundstock has survived.

TWO miles east of Poundstock lies **Penfound Manor**. The Revd Sabine Baring-Gould, vicar of Lew Trenchard just across the Devon border, wrote in *An Old English Home* (1898) 'in a dip in the land, at the source of a little stream, snuggling into the folds of the down, bedded in foliage, open to the sun, hummed about by bees, twinkled over by butterflies, lies this lovely old house', the home of the Penfound family from the early fourteenth to the mid nineteenth century.

The earliest official record of the manor is in the Domesday Book, when it belonged to William the Conqueror's half-brother Robert, Count of Mortain. Called Penfou in Domesday (meaning head or source of the stream) it was known as Penfound by the sixteenth century. It is not known exactly when the de Penfounds took over the manor, but documents indicate that they were in residence in the reign of Edward III.*

As the Penfounds grew richer, so the house increased in size. In early Norman times it consisted merely of the great hall, two walls of which remain, more than six feet thick. The existing chimney and fireplace were probably added in the early thirteenth century. The scarlet and blue glass in the border of the Elizabethan window (to the left of the porch) came from Westminster Abbey after the second World War, when fragments of Abbey glass were made available for restoration work.

The first addition to the house was the Norman wing built at right angles to the south-west corner of the great hall. It consisted of a withdrawing-room for the ladies downstairs and solar above, reached originally by a stone spiral staircase. The women of the household could sit by the large mullioned window (which still retains its original glass) and enjoy the sun in this south-facing wing. The solar was the only bedroom until 1589, when the main staircase was built and a second bedroom added over the mediæval buttery, the first room to be built to the east of the screens passage. While the great families abandoned their fortified castles for magnificent mansions during this period of rising prosperity, country squires made their modest manors more comfortable, better heated, and better lit. Penfound was no exception.

No further building took place until 1635, when Arthur Penfound added a dairy and a new kitchen with bedrooms over each. He also laid a pebble path which runs from the porch, through the screens passage, to beyond the back door. He incorporated the date, 1638, in white pebbles at the entrance. Horses could now be taken straight through the house to be watered at the horse-trough over the well, instead of having to be led round the outside of the courtyard. The spandrels of the carved granite archway bear the initials of Arthur and his wife Sibella. The inscription 'In the Yeare 1642' is carved across the lintel. Arthur's son Thomas, a royalist like his father, planted a Judas tree in the courtyard on 30 January 1649 to record his hatred of Charles I's executioners. The tree was struck by lightning in the 1930s and a sundial now stands in its place. Another Judas tree has been planted near the entrance gateway.

Penfound Manor, Poundstock. The entrance porch bears the inscription 'In the Years 1642' and the initials of Arthur and Sibella Penfound.

The Penfounds never knew great wealth – perhaps they were not grand enough to attract an heiress, nor sufficiently ambitious to seek distinction outide Cornwall – which accounts for the modesty of the successive alterations. Their fortunes went up and down, as happens to most families, and they suffered financially for their support of the Pretender in 1715. The last Penfound died in the Poundstock poor-house in 1847 'leaving issue in the state of poverty'. Fortunately, no subsequent owner enlarged the house, as so many Victorians were wont to do, and it remains an outstanding example of a mediæval manor house with sixteenth- and seventeenth-century improvements. Looking through the wrought-iron gates into the quiet courtyard, now gay with geraniums and pink hydrangeas and down the path to the house itself, with its uneven slate roof and welcoming porch through which horses once passed, it is as if time has stood still at Penfound.

4.

West

And have they fixed the where and when
And shall Trelawny die?
There's twenty thousand Cornish men
Will know the reason why.

R. S. Hawker

THREE rivers run the length of West Hundred, a small area sandwiched between East Hundred and Powder Hundred on the south coast. The river Fowey forms a natural boundary to the west, the river Seaton to the east. In between flow the river Looe and its tributaries. These three lushly-wooded river valleys cut through otherwise barren countryside, and great estates grew up in or near them.

Glynn stands high on a hill above the river Fowey, not far from Liskeard. Glynn is a Celtic word meaning valley, and the family who settled here soon after the Conquest took the same name. Centuries later, in 1805, Edmond John Glynn built a massive square late Georgian mansion on the site of the old one.

The heavy eaves and small windows (compared with those of Georgian houses elsewhere in England) of the south front show the architect's desire to protect this exposed side of the house, which looks right down the valley. On a rainy day the mist swirls up the hillside like some silent, stealthy spirit come to envelop the house in a damp grey shroud; when it is fine the silvery granite sparkles in the sunlight and you can see down the parkland to the wooded valley bottom and into the country far beyond. Such a sense of space round the house, a rarity in a county where shelter is the first consideration, adds to the grandeur of Glynn. The elegant windows of the entrance front with their crisply carved urns look very graceful from a distance.

Edmond John Glynn lived in this magnificent mansion for only a few years. According to Gilbert in his *Parochial History of Cornwall* (1838), his uncle had him proclaimed a lunatic - he apparently spoke to no one but communicated his thoughts by writing - and thereby inherited Glynn. But his ill-gotten gain was not enjoyed for long. On a chilly November day in 1819 a fearful fire broke out which swept through the main block. A brew-house at the back of the house escaped the furnace, but otherwise only the external walls survived. It was this huge stone shell which a dashing young cavalry officer bought in 1825.

Sir Hussey Vivian had returned victorious from Waterloo with numerous military decorations. He was determined that his new home should celebrate his achievement and honour his name. The portico of the entrance front and the giant columns of granite of the west front, otherwise built of cream-coloured Bodmin stone and part of the earlier house, are probably Sir Hussey's architectural contribution.

The west front looked on to a formal Italian garden with a fountain in the middle. A conservatory and octagonal pavilion were added to the

Glynn, near Cardinham. The house was gutted by fire in 1819 and remodelled by Sir Hussey Vivian soon afterwards.

Glynn. The drawing-room window to the left of the porticoed entrance.

façade during the Regency period. The drawing-room, little drawing-room and dining-room were all on this side of the house, their lofty ceilings decorated in the centre with beige and gold replicas in plaster of Sir Hussey's medals and orders. Although the Italian garden has long since vanished, the huge crimson rhododendrons and azaleas of the American garden, a large shrubbery on the slope above the formal flowerbeds, are still a sensational sight.

The Vivian family remained at Glynn until 1947, when the estate was sold. Neglected for over twenty years, the house was saved from demolition by its present owner who has restored it with great care. The glass roofs of the conservatory and pavilion, the latter a 'museum' where curios were displayed in the Vivians' day, were sensibly removed but the stonework preserved. Now a pink clematis trails over the remaining structure and plants replace copper pans on the stone shelves of the dairy behind.

A FEW miles from Lostwithiel and just south of Glynn lies another great estate. **Boconnoc** was the seat of the Carminows from very early times, but passed to the Courtenay family when a Carminow daughter married Sir Hugh Courtenay some time in the fifteenth century. Henry VII bestowed the title of Earl of Devon on Sir Hugh's son, Edward Courtenay - the title was forfeited when Thomas Courtenay, Earl of Devon, was beheaded for his Lancastrian sympathies after the battle of

Glynn. The remains of the octagonal pavilion and conservatory which once overlooked the Italian Garden.

Tewkesbury in 1471. Edward's son married, as Henry VII had done, a daughter of Edward IV, so that his heir, created Marquis of Exeter, was a first cousin of Henry VIII and a potential claimant to the throne. Not surprisingly, he was arrested in 1538 and executed on a trumped up charge of treason. All his estates became vested in the Crown, including Boconnoc, which was not long afterwards granted to the Russell family.

In 1579 the Russells sold Boconnoc to William Mohun whose descendants lived there for over a hundred years until in 1712 Sir William's great grandson, Charles, was killed in a duel. *

Five years later, Charles Mohun's widow sold Boconnoc to the late Governor of Madras, Thomas Pitt, who raised the purchase price by selling the fabulous Pitt Diamond which he had bought from a native in 1702. This was sold to the Regent of France, Philippe D'Orleans, for the princely sum of £125,000 and was eventually set in the hilt of Napoleon's sword. Pitt remodelled Boconnoc House, adding a new wing to the old structure. This east wing, with its elegant neo-classical dining-room, adorned with shell niches and pedimented doors, looked towards the little church of Boconnoc which stands on a bank just above the house.

Governor Pitt's great-grandson, the 1st Baron Camelford, added a further south wing overlooking the vast expanse of lawn and adjoining parkland. This wing contained an immense gallery on the upper floor where hung a magnificent collection of portraits. Lord Camelford was

Boconnoc, near Lostwithiel. The east front.

Boconnoc. View from the south front, showing part of the six-mile-long drive laid out by the first Lord Camelford.

Boconnoc. A picturesque waterfall in the woods.

also a keen landscape gardener and an early enthusiast of the picturesque. He laid out a six mile drive through the oak and beech woods beyond the park, with sudden vistas of waterfalls, quaint bridges and grottos. The little river Lerryn, a tributary of the river Fowey fed by numerous streams, curls its way through the grounds so that the sound of running water is never far away. Today, the massive clumps of rhododendrons in the woods are an unforgettable sight in May.

High on a hill behind the house is an obelisk 123 feet tall, erected in 1771, in memory of Lord Camelford's uncle Sir Richard Lyttelton. It stands in the middle of an old military entrenchment that goes back to the Civil War. ★

Walking through Boconnoc woods, home of the buzzard and the badger, and up to Boconnoc House high on its knoll, it is comforting to think that King Charles, whose fortunes were so soon to turn, enjoyed a month of safety here. Even nature was in sympathy with him. Legend has it that while King Charles was receiving the sacrament beneath the boughs of an old oak tree on which his standard was displayed, a shot was fired which narrowly missed him. The bullet struck the oak, which for ever afterwards grew in the most stunted fashion and sprouted mottled leaves. This oak was cut down by the uncle of the present owner, but another has been planted in its place.

In 1804 Lord Camelford's son Thomas was killed in a senseless duel in London at the age of twenty-nine. ★

Boconnoc passed to Thomas's elder sister Anne Pitt. She had no children and it then went to her husband's nephew, George Matthew Fortescue, of Castle Hill in North Devon. Fortescues have lived at Boconnoc ever since, although no longer in the big house which is now empty.

As at Boconnoc, **Morval** house and church stand close together, tucked out of sight down a sleepy lane on the outskirts of Morval village, two miles north of Looe. In days gone by a boat would have brought a visitor up the River Looe as far as the lock about a mile from the house, which stands at the head of rich woodland further up the valley. A stream flows through the woods and widens into a small lake or inlet before joining the river, so that the grounds have great natural beauty. Trees on a hillside behind the house give shelter while on the other side grassy slopes stretch down to undulating parkland.

Morval originally belonged to the Glynn family, evidently relatives of the Glynns of Glynn since both families bore the same coat of arms. Feuds were common in Tudor times and Morval was the scene of a particularly vicious one, which led to the murder of John Glynn. ★★

Despite this tragedy the Glynns lived on at Morval until well into the sixteenth century, when Thomasine Glynn brought the manor in marriage to Richard Coode. In 1637 John Buller, younger son of the

Morval house and church, near Looe. 19th-century engraving.

MP for Shillingham in Devon, married the last of the Coodes. He and his descendants occupied Morval for 250 years. His son John Francis increased the family's wealth and standing when he inherited Shillingham from his elder brother.

Retaining the shape of the original E-shaped block, Morval is a pleasing two-storied house with large four-light mullioned windows and simple hoods mouldings. It was built largely in the seventeenth century, when the Bullers came to live there, but the white criss-cross glazing bars were inserted in the eighteenth century and the hipped roof, too, probably belongs to this later period. A low, buttressed wall bordered by flowerbeds enclosed the terraced walk along the front of the house in a neat rectangle, the two slightly projecting wings providing additional shelter and privacy. To the right of the house stands the small, low church of St Wenna, half hidden behind the rhododendrons, where Glynns, Coodes and Bullers lie buried.

The last of the Bullers died childless in 1890, and Morval passed through the female line to the Tremayne family. It has been well maintained ever since.

THE manor of **Trelawne**, just outside the village of Pelynt, to the west of Looe, was once the home of one of Cornwall's most illustrious families, the Trelawnys. The house, now a country club, has survived but the garden has been transformed into an amusement centre. Hard black tarmac has replaced the gravel paths and lawns that once surrounded Trelawne, its tranquil parkland now a noisy landscape of caravans, swimming-pools, children's slides and swings.

William Bonville lived at Trelawne in the fifteenth century, and the castellated towers of the semi-fortified Tudor dwelling can be seen on the eastern side. The rest of the house was rebuilt by Sir Jonathan Trelawny when he bought Trelawne in about 1600. A century later a chapel was erected on the south side, on the site of the old one, and a Georgian block with sash windows was added to the mediæval structure. ★

Bishop Trelawny, the 3rd baronet, was by far the most colourful member of the family at Trelawney. He was a jovial character who fathered eleven children and had a reputation for swearing ('I do not swear as a priest but as a baronet and a country gentleman'). It is assumed that he inspired the poem with its rousing chorus 'And have they fixed the where and when and shall Trelawny die, there's twenty thousand Cornish men shall know the reason why', revived by Robert Stephen Hawker and now almost the national anthem of Cornwall. ★★

During the Trelawny's long reign at Trelawne – they remained there until the beginning of this century – the West Looe river which threads its way through what was then Trelawny territory was an important waterway. The tide brought up plump salmon and gleaming sea-trout which were caught in a weir known as Shallow Pool and served at the baronets' dinner table. †

The Trelawne estate was broken up in the 1920s and the house became for a time a home for retired Church of England clergymen. Threatened with demolition in the 1950s, it is fortunate to have survived. The new owners of Trelawne were willing to meet the demands of a new age in which leisure, not industry, forms the basis of Cornwall's economy.

Trelawne Manor, Pelynt. 19th century engraving showing the Tudor castellated towers of the east side and the chapel built on the site of the old one c. 1700.

5.
Trigg

Aside from the town, [Bodmin], towards the north sea,
extendeth a fruitful vein of land comprising certain parishes,
which serveth better than any other place in Cornwall for winter
feeding, and suitably enricheth farmers. Herethrough, sundry
gentlemen have there planted their seats. . .

Richard Carew, *The Survey of Cornwall*, 1602

CAREW did not have a good word to say about Bodmin, deeming it to
be the most 'contagiously seated' town in the county. The front rooms
of the houses on the main street were dark and cold because a high hill
blocked out the sun on the south side, while 'the back houses, of more
necessary than cleanly service, as kitchens, stables, etc., are climbed up
unto by steps, and their filth by every great shower washed down
through their houses into the streets'. Nonetheless, Bodmin was a large
and wealthy town in the Middle Ages, having been important as a
religious community ever since St Petroc founded a monastery there in
the sixth century.

Since very early times, traders from Ireland afraid of the treacherous
seas round Land's End would put into the port of Padstow and sail their
boats up the Camel estuary to Wadebridge and beyond. They were
then within striking distance of Bodmin. Cargoes of white fish, coarse
Irish cloth for mantles, and prepared timber were taken the
comparatively short distance overland to Lostwithiel, situated at the
top of the Fowey estuary on the south coast. From there, the goods
went by ship to Brittany. In return for Cornish tin, hides, herrings and
pilchards, the Breton boats sailed into Fowey filled with salt, linen,
woollen cloth, canvas and, most valuable of all imports, wine.

Situated midway between Padstow and Fowey, Bodmin benefited
handsomely from its position along this trade route. Foreign merchants
and traders attended its half-yearly fairs which, unlike ordinary
markets, were festive occasions designed to draw people from far and
wide. They offered a chance to see and buy foreign wares and to hear
news from other parts of the county and indeed from England. Much
later, Bodmin became a kind of central link in the road system. In the
eighteenth century, stage coaches travelling east-west on the great mail
road from Launceston to Truro and Falmouth or north-south from
Camelford to Truro all stopped at Bodmin, which thus became an
important crossroad.

Small wonder, then, that 'sundry gentlemen', to use Carew's words,
chose to live near this prosperous and lively town where good
conversation could be had and where friends from other parts of
Cornwall frequently met. They could farm the fertile pockets of land
towards the coast, involve themselves in county affairs, legal and
administrative (the quarter sessions were held in Bodmin), and journey
to London with comparative ease.

Bokelly, near St Kew. A Tudor buttressed barn.

William Carnsew, who lived at **Bokelly** near St Kew in the latter half of the sixteenth century, was one such gentleman. A Tudor barn with hefty buttresses still exists, and a house too, although this was re-fronted towards the end of the seventeenth century. The view across the stubble fields to the skyline is the same now as it was in Carnsew's

Bokelly, re-fronted towards the end of the 17th century.

time. No other buildings have been erected, and a windy track still leads down to the house which is sheltered by a small copse.

According to Carew, William Carnsew was 'a gentleman of good quality, discretion and learning'. Surviving fragments of his diary for the year 1576-7 give a vivid picture of life at Bokelly. His preoccupations are those of an intelligent country squire, content to farm his land and visit his friends. Like any keen farmer, he notes weather conditions, takes an interest in sheep-shearing and the selling of wool, worries about his harvest of oats and wheat and the cutting of his meadows; like any Cornishman, he hears the wind and watches the sea. Three ships are cast on to the rocks, he reports, in a matter of a few days that stormy November.

In his spare time this jovial, gregarious man plays bowls, quoits, and saint with the locals, sets off on horseback along narrow lanes to stay with the Arundells at Lanherne or Trerice, or merely gallops across the fields to his neighbours the Roscarrocks at nearby St Endellion. Other friends lodge with him at Bokelly on their way to the assizes - Sir John Killigrew, for example, to whom Carnsew lends a fresh horse - or come to make merry on Twelfth Night. 'Many hawked with me,' he writes; 'George Grenville, Richard Carew, Richard Champernowne, their brethren.' They discussed current affairs as well as local matters over good food and wine. The Protestant and Catholic reaction to the Papal Bull that excommunicated Elizabeth I might have been one topic of lively conversation, for Carnsew's friends were from both camps. *

Old Carnsew was a devoted husband and father. His three sons were all at Oxford in the 1570s, the youngest, William, being made a Fellow of All Souls in 1579. They were a close-knit happy family. William wrote affectionate letters to his brother Richard at Bokelly from his chambers in London, hinting that he was finding it hard to make ends meet. His father visited him, bringing him money and, to his delight, books. His mother sent him new shirts, wondering as she sewed on the butons, whether his landlady was feeding him adequately.

Young William was Member of Parliament for Camelford in 1597-8 and 1601. He relished the contact this gave him with Cornish folk and their affairs. In 1610 he married Ann Arundell of Trerice, a match that brought great pride and joy to his father who knew this respected family so well. But neither William nor his brothers Richard and Matthew produced children, and the Carnsew line came to an end. Bokelly passed by sale to the Tregagle family in the mid seventeenth century and ultimately became a Molesworth property. It is father Carnsew's diary, however, that brings Bokelly to life, and his spirit wanders there today.

A FEW miles to the north-west of Bokelly, between St Endellion and the coast, lies **Roscarrock**, which (according to Carew) 'in Cornish meaneth a flower, and a rock in English'. Little remains of the Tudor house that William Carnsew and Richard Carew used to visit, but among the outbuildings of the present late Georgian farmhouse is a range which testifies to the importance of Roscarrock. Above a secondary room on the ground floor is a large upper room with a four-light oriel window and an exceptionally handsome timbered roof. This must have been either a first-floor hall or a solar.

The history of Roscarrock is a sad yet noble one. The family were staunch Roman Catholics prepared to suffer for their faith. In 1577 Nicholas Roscarrock, a barrister and antiquary, was accused at Launceston assizes of failure to attend Protestant services; three years later he was imprisoned in the Tower and was not released until 1586. He joined Francis Tregian, another brave Cornish recusant, in the Fleet prison and whiled away the time by compiling an account of the British saints, translating the lives of a number of Cornish saints from the Cornish with which he was conversant.

Known adherence to Rome meant not only imprisonment but the risk of losing one's lands and possessions. Moreover, virtually no Catholic family, with the exception of the wealthy Arundells of Lanherne, could afford to pay the fine exacted by the government in the 1580s for non-attendance at church. The Roscarrocks lost everything, and their impoverished estate passed into the hands of the Crown. Nicholas never saw Cornwall again, and spent the last years of his life at Naworth with Lord William Howard. Distressed as they were to see him endure such hardships, his Protestant friends such as Richard Carew, who praises him most warmly in his *Survey* for 'his industrious delight in matters of history and antiquity', secretly admired his courage and the strength of his conviction.

TRESUNGERS, another house in St Endellion parish, belonged in Nicholas Roscarrock's time to the Matthews family. Unfortunately for us, Carew does not mention Tresungers in his *Survey*, but *Mathew* is marked clearly as the owner on Norden's map. This late sixteenth-century farmhouse is L-shaped and has a gabled wing. The castellated three-storey entrance tower, ivy-clad, is dated 1660, and was doubtless built to commemorate the Restoration of Charles II. Tower porches such as these are a feature of many other Cornish houses of the period.

A MILE to the north of Bokelly, just outside the hamlet of Trelill, stands **Pengenna**, a fine seventeenth-century farmhouse. It is said to have been built by the Revd Thomas Pocock, whose initials are carved over the entrance. Like so many other well-to-do farmers, the Pococks found the mediæval house dark, uncomfortable and constricting, and decided to build a new, spacious house against the old Tudor block. Built of the local rubblestone, it towers above the earlier house and looks exceedingly handsome with its mullioned windows and hoodmoulds.

The charm of Pengenna lies in the fact that it is built at right angles to the steeply-sloping land at one side. You approach it from behind. The rear elevation, with its tiny attic windows above the first floor and its stalwart buttress, seems on the defensive and definitely forbidding. But as you walk past the Tudor block and turn into the courtyard, the smiling face of the front of the house bids you welcome, its Delabole slate roof glistening in the sun. Fine old outbuildings are clustered round the farmyard, from which you can look south down the wooded valley to Tremeer.

Pengenna. The old Tudor block.

Pengenna, Trelill. The 17th-century entrance front built to the right of the old Tudor block.

TREMEER GARDENS, just outside St Tudy, are a twentieth-century creation of outstanding beauty and of interest to any horticulturist. Eric Harrison, soldier and veteran of the First World War, bought Tremeer in a semi-derelict condition in 1939. More interested in hunting than in gardening – he was master of the North Cornwall hounds – it was not until after the war, when he retired from the army, that he devoted any attention to the garden. In 1946 Tremeer boasted only one magnolia, a few rhododendrons choked by nettles and brambles, and the remains of a herbaceous border. Some mature, deciduous trees gave shelter but that was about all. With the eye of an aesthete, Harrison realised the potential of the seven acres of ground he had acquired and set about transforming it.

The success of the overall design is immediately apparent. Nothing seems contrived, and yet wherever one strolls – down the gently sloping lawns towards the two ponds and mill-leat or across the grass to left or right – pleasing vistas and wonderful colour greet the eye. Evergreen oaks, *Cupressocyparis leylandii thuya* and *Griselinia littoralis*, as well as conifers and sweet bay, break up the broad swathe of lawn in a natural way, giving a sense of intimacy and variety. The trees are so placed that a number of different routes to the bottom of the garden are introduced. Walk straight down the middle of the lawn to an old oak tree – the ground beneath it a mass of wild cyclamen in September – or wander in a horizontal or diagonal direction across the grass into small clearings created by the trees. Whichever way you go, the bottom of the garden is a treasure-trove. After the subtle greens, pale and dark, a gawdy array of reds and pinks startles the visitor. Rhododendrons were Harrison's great love and he planted them in profusion. He learned how to grow them from seed or from cuttings and experimented with hybridisation quite early on. In fact he was so successful that the Royal Horticultural Society awarded him many prizes over the years. In 1961 Harrison married Roza Stevenson, the owner of Tower Court, Ascot, famous for its rhododendrons. Two hundred or so unwanted shrubs were shifted to Tremeer, where they now flourish magnificently. Nearly all are ticketed so that gardening enthusiasts can learn as they marvel at the achievement of one man and his gardener.

Harrison despaired of success because honey fungus, particularly virulent in the soft, moist Cornish air, attacked and killed many of the rhododendrons he planted. He turned to camellias as an alternative, planting them at random along the grassy paths that lead to the pond. In early spring this nonchalantly-planned shrubbery is a breath-taking sight. The rhododendrons and camellias almost overhang the pond where ducks waddle and quack. It is shadier, more enclosed here; shafts of light through the trees dapple the water on a sunny day, and crimson petals make bright dots of colour on the surface as they float slowly by. Candelabra primulas, irises and moisture-loving perennials grow on the banks of the pond. Swans and deer made of lead are silent

Tremeer Gardens, St Tudy. View from the terrace.

Tremeer Gardens.

Tremeer, rebuilt in 1798 with an upper storey added c. 1900.

companions at the water's edge, guardians of the beauty around them.

As in most Cornish gardens devoted to camellias and rhododendrons, late summer colour is a problem. At Tremeer the solution is a wide border flanking the gravel path along one side of the house. This is filled with hydrangeas, eucryphias, enkianthus and embothrium, hypericums and heathers, which flower well into the autumn.

In the early fifteenth century the manor of Tremeer belonged to John Treffry. By 1570 it had passed to Edward Lower whose mother was a Treffry, and the Lower family was to remain here for more than a hundred years. Doctor Lower, Charles II's physician, was born at Tremeer and was rumoured to have carried out experimental blood transfusions on rats and dogs on the premises. Tremeer House was rebuilt in 1798 and altered extensively in about 1900 when an extra storey was added, which rather throws it out of proportion. The semi-circular bay windows of the garden front provide glorious views not only of the garden but of the rolling fields beyond. In the eighteenth and nineteenth centuries the house changed hands several times, but did not benefit in any way until Harrison arrived in 1939. Due to his imagination and expertise, and the present owners' hard work, Tremeer Gardens are superbly kept and lovely to behold.

PENCARROW, the grandest house in Trigg Hundred, lies just north of Bodmin at the head of a thickly wooded valley through which the river Camel flows. Pencarrow belonged long ago to the ancient families of Stapleton and Sergiaux and is thought to have passed from them to a

family who took their name from it, the de Pencarrows. In 1497 a
Pencarrow joined other Cornishmen in an ill-fated uprising against an
unfair levy on tin. Branded as a traitor, he conveyed the estate to a
neighbouring nobleman. Lord Marney of Colquite (near St Mabyn –
the ruin of a late fifteenth-century house still exists below the present
one), hoping that his influence might save his life. Pencarrow then
passed to Lord Marney's heirs, an Exeter family named Walker, who
sold it to John Molesworth during the reign of Queen Elizabeth. The
Molesworth-St Aubyn family who own Pencarrow today are his
descendants.*

The present house, an imposing, square Palladian-style mansion
built in the 1760s by the 4th and 5th Sir John Molesworths stands at the
end of a mile-long drive which is one of the loveliest in Cornwall. The
visitor proceeds first past an Iron Age fortified encampment whose
inner ditch is a perfect oval, and then into a wood of tall, stately beeches
and old oaks, the sunlight glinting through their branches, and making
flickering shadows on the ground. Banks of deep cobalt-blue
hydrangeas line the last part of the drive, which finally curves sharply
round to the east front of the house.

After the natural woodland setting of the drive, the elegantly laid-out
garden round two sides of the house comes as a surprise. A garden in
Cornwall which even hints at formality is something of a rarity since so
many are wild, romantic gardens carefully cultivated to appear as if
they had never been touched by human hand.

Humphry Repton would have approved of Pencarrow's terraced
flowerbeds round the house, its huge circular lawn, fountain and neat
gravel paths. A garden, he said, 'is a piece of ground fenced off from
cattle and appropriated to the use and pleasure of man: it is or ought to
be, cultivated and enriched by art, with such products as are not natural
to this country'. At Pencarrow a straight gravel path leads from the
central fountain and up two sets of steps to the parkland fence, beyond
which cattle graze. The manicured lawn is surrounded, on a higher
level, by mature shrubs set in a rock garden, the granite boulders
heaved from Bodmin Moor being the only ornaments among the
rhododendrons and camellias. Trees have been carefully placed so as
not to interrupt the vistas to the house, and blend naturally with the
park scenery and woodland beyond.

The interior of Pencarrow reflects the tastes of its owners over the
past two hundred years. Friendly and knowledgeable guides, who
know the house and its history intimately, bring the rooms to life. The
tour begins in the entrance hall, turned into a pine-panelled library by
the 8th baronet in the first half of the nineteenth century. A display of
porcelain, European and Chinese, deserves a close look, and in the
same cabinet are a set of beautiful Georgian wine glasses with airtwist
decoration in the stem, as well as an unusual collection of glass ink pens
blown for the 1851 Great Exhibition.

Pencarrow, near Bodmin. The entrance front c. 1760.

The Molesworths were travellers who appreciated French and Italian modes of decoration. A canapé by Georges Jacob, with its original tapestry of roses, stands between the windows in the drawing-room, and the furniture is gilded in the French manner. The chairs and curtains are covered in deep musky pink flowered silk damask which was taken from a Spanish galleon captured off the Philippines. It was on its way to the Argentine with a cargo of Chinese silks. The admiral responsible for the operation was a relative of the 6th baronet who brought the silk back as a present. The dark *gros bleu* Sèvres plates and candelabra set on the white marble Adam chimney-piece show how effective a mixture of styles can be.

The rococo plasterwork of the music room ceiling is wonderfully light and airy. Its four corners depict the four seasons - daffodils for spring, corn for summer, grapes for autumn and fire for winter. The shell, bird, and floral motifs of the overall design are so graceful that one suspects a French or Italian craftsman was at work.

A liking for the oriental is also evident, in fine examples of Chinese porcelain of the Kang Hsi period, and there is a rare Chinese cabinet in the drawing-room. Its interior is designed to look like a stage with tiny actors on it. The walls of an ante-room off the dining-room are covered in eighteenth-century hand-painted Chinese linen with a beautiful bird design in soft blues. A large bowl on a table depicts Pencarrow through Chinese eyes. Drawings of Pencarrow were sent over to China for the artists to copy, but they turned the house into a Palladian pagoda and the horses and hounds in front have decidedly slanted eyes!

Family portraits by Northcote, Raeburn, Beechey and Reynolds adorn the walls upstairs and down, and a moving portrait of Charles I seated at his trial hangs on the upper staircase. It is painted by Edward Bower, who probably made sketches of the King at the trial before painting four variants of the monarch. It is interesting to compare this version with the one already mentioned at Antony.

Upstairs, Pencarrow has a more informal air. Regency and Victorian furniture appears in the bedrooms, which all look lived-in; wide-eyed, prettily dressed Victorian dolls are propped up on the pillows. The Molesworths seem to have held the view that 'one should hang on to everything', with the happy result that a dolls' house, perambulators, toys and dolls survive as souvenirs of childhood a century ago.

A MODEST, two-storeyed rectangular manor house called **Kirland** stands on a hill surrounded by tall, spindly trees full of cawing rooks on the outskirts of Bodmin. It is remembered fondly by Emma Gifford, Thomas Hardy's first wife, whose home it was in the 1860s. 'This little old manor-house has some good points about it,' she writes in her *Recollections*, which Hardy found after her death.★

Its stuccoed walls are painted a warm beige, and from the porch one

Kirland, Bodmin. The home of Thomas Hardy's first wife, Emma Gifford.

looks down the sloping lawn to the stream and fields below and beyond. It has changed little in a hundred years.

6.

Pydar

Pydar in Cornish is Four in English, and this is the fourth hundred of Cornwall if you begin your reckoning from the western part, at Penwith.

Richard Carew, *The Survey of Cornwall*, 1602

ALTHOUGH Norden dismissed Pydar as a barren region of heath and turf that 'wanteth woode', it boasts some fine houses with interesting histories. Pydar is rich in Celtic memories because of its association with St Petroc, who founded a monastery on the southern banks of the river Camel in the sixth century - the site of Padstow today. Fierce Danish raiders ravaged Padstow in 981 and St Petroc's relics were moved to Bodmin, a safer inland site. The earliest record of Bodmin Priory dates from the mid eleventh century, and for the next five centuries, until the dissolution of the monasteries in 1536, it controlled vast estates in north Cornwall, including property in the Padstow area.

Rialton, which at first glance looks nothing more than a commonplace Cornish stone cottage, slightly larger than most, stands just off a narrow road outside St Columb Minor. The front of Rialton faces away from the road, but as soon as the visitor enters the little courtyard, it becomes clear that this is no ordinary dwelling but the main wing of a late fifteenth-century monastic manor, comprising a hall and kitchen with a solar above. The manor of Rialton is mentioned in Domesday when it had seven hides of land. The canons kept two hides and the villeins had the remaining five. There were also 60 acres of wood and 300 of pasture. In the early sixteenth century the priors of Bodmin decided to turn Rialton into a pleasant country residence. Prior Vyvyan, the last but one prior before the dissolution, added the wide porch with the study above. In two of the study's three six-light mullioned windows are fragments of glass with the initials TV - Thomas Vyvyan - and three fishes, the arms of the priory. Next to the study is the prior's oratory or bedchamber, which would once have had a fine view over the estate's meadows and orchards.

Prior Vyvyan died in 1533 and was succeeded by Thomas Mundy.★

Five generations of Mundys lived at Rialton until Parliamentarians seized the property during the Commonwealth. At the Restoration the Godolphins were granted the lease but never lived there, although the 1st Earl Godolphin took Viscount Rialton as a second title. Rialton became a farm and, in about 1870, when the Duchy built a new farmhouse further up the valley called Rialton Barton, became known as Rialton Mill. The east or service wing at right angles to the remaining one was then pulled down.

NICHOLAS PRIDEAUX was Prior Vyvyan's man of affairs and this association served him well. Acting upon the dying prior's wishes, he ensured that Secretary Cromwell appointed Mundy as Vyvyan's

Prideaux Place. Engraving from Borlase's Natural History of Cornwall (1758) showing the monuments erected in the grounds by Edmund Prideaux in the mid-18th century.

successor. Nicholas was duly rewarded by Mundy with the lease of the great tithes of the four parishes of Egloshayle, St Minver, St Cubert and Padstow. Tithes, mainly from churches whose rectories were appropriated to them, were the other chief source of monastic income, but these, like the revenues from ecclesiastical estates, fell into the hands of the laity in the form of long leases which the Crown found impossible to repossess after the dissolution. Near the tithe–barn of the monks at Padstow, 'a new and stately house' according to Carew, was built by Nicholas Prideaux's great-nephew in 1592. It was called Place. 'Place' means palace, and other houses in Cornwall that were built on monastic property, such as Place at Fowey, bear the same name.

Prideaux Place, as the house is now called, is surrounded on three sides by belts of trees planted in the eighteenth century. The fourth side, or entrance front, looks across the deer park to the high ground on the other side of the Camel estuary. This long east front, with its three projecting bays, forms the characteristic Elizabethan E shape. Edmund Prideaux, son of the Dean of Norwich, who inherited Place in 1728, probably altered the façade slightly. A drawing of his shows that the projecting bays originally had pointed gables. It seems likely that it was he who removed these and extended the castellations in the horizontal line we see today.

The south front has been subject to many more changes, since the principal rooms, always the first to be altered to follow new fashions, are on this side. Another drawing of the 1730s by Edmund Prideaux, made before any alterations took place, shows that this side of the house was a simple two-storeyed rectangular block with five irregularly placed bays. This lack of symmetry no doubt offended the Georgian Edmund, who left another drawing with a scale at the bottom, presumably as his scheme for alteration. He retains the shape, but inserts two more bays – his windows are no longer two-light mullions but elegant sash windows – and continues the two string courses right the way along the façade to create a more classical effect. In 1810 the

Prideaux Place. 19th century engraving showing the square library tower and round drawing-room bow of the south front, added c. 1810 by the Rev. Charles Prideaux-Brune.

Revd Charles Prideaux-Brune added the square library tower at the left end of the façade and the round central tower or large bow of the drawing-room and principal bedroom above. The deliberate irregularity and decorativeness of the 'Gothick' windows and bobbly pinnacles create an informal, picturesque effect similar to that which Nash achieved at Luscombe in Devon ten years earlier. Since the removal of the top layer of the bow and all the pinnacles twenty years ago (they were no longer safe) the south front has lost something of its charm and looks more uniform than intended. The delicate 'Gothick' plasterwork of the drawing-room, library and hall ceilings remains. The late seventeenth-century panelling and carvings in the Reading Room came from the demolished Grenville seat of Stowe. There is a painting of Sir Bevil Grenville whose daughter married into the Prideaux family.

Despite such alterations to the original interior, the Elizabethan lay-out can still be discerned. The entrance porch on the east front led into a screens passage with a single-storey great hall to the south. This was redecorated in the eighteenth century, presumably by Edmund, and the original screen removed. The existing screen, with its superbly inlaid frieze and panels of fantastic animals and floral designs (thought to be Spanish work of the late sixteenth or early seventeenth century), was probably fitted when the hall was turned into a dining-room in the nineteenth century. The great chamber, above the great hall, was split up into a number of separate bedrooms at the same time. The original barrel-vaulted plaster ceiling, until recently concealed beneath a lower one, is well preserved. The delicately modelled panels depict biblical scenes and characters including Susannah and the Elders and are linked together by intricate strapwork and smaller panels framing birds and

St Benet's, Lanivet. The windows of the 15th-century Benedictine foundation are incorporated into the 19th-century façade, behind which stands the ruin of the old church tower.

animals. The gallery at Lanhydrock, to be discussed later in this chapter, has a plasterwork ceiling of the same quality and style and clearly shows the same team of plasterers at work.

Edmund Prideaux was not only an amateur architect and interior decorator but also a keen landscape gardener. Following contemporary custom, he erected a number of small monuments in the grounds, as Sir William Chambers did at Kew at much the same time. A view of Prideaux Place in Borlase's *Natural History of Cornwall* (1758) shows an Ionic temple behind the house and an obelisk and small monument to the south side. The obelisk has gone but the two other buildings survive, though now hidden by trees. A castellated gateway flanked on either side by a low battlemented wall also appears in Borlase's view and forms the entrance today.

A THIRD house in Pydar with monastic rigins is **St Benet's**, which can be seen from the road just outside Lanivet a few miles south west of Bodmin. An early fifteenth-century Benedictine foundation, originally a lazar house or hospital, it passed to the Courtenay family at the time of the Dissolution. Today, its fifteenth-century windows are incorporated into the mid-nineteenth-century façade. The two niches and small oriel window with their ogee heads also come from the original building. The tower of the original church can be seen at the back of the house.

Thomas Hardy's first wife, Emma Gifford, has left us a picture of St Benet's in her *Recollections*.★

The house deteriorated rapidly after Emma's time, but was restored by enterprising owners in the 1960s. It now looks beautifully kept, with an attractive garden in front. The old tower behind has rightly been kept a ruin, although the top storey has been altered, to remind us of its history.

THE beautiful Vale of Lanherne, the only richly wooded valley in Pydar, runs down to the sea at Mawgan Porth, half-way between Padstow and Newquay. It once belonged to the richest and most powerful family in Cornwall, the recusant Catholic Arundells of **Lanherne**. Just above the church in the village of St Mawgan, you can see the high walls that surrounded their home. ★

Lanherne fell into decay in the early eighteenth century when the heiress of Lanherne married the 7th Lord Arundell of Wardour and left Cornwall. However in 1794 Henry, 8th Lord Arundell, who was responsible for the building of Wardour Castle as we know it today, generously assigned Lanherne to a group of English nuns who, upon the French invasion of Belgium, had been forced to leave Antwerp. The sisterhood has flourished here ever since, for Lanherne remains a convent to this day.

One can only glimpse the top of the late seventeenth- and eighteenth-century parts at the back of the house from the other side of the high walls, but the Elizabethan entrance front can be approached from the little road that runs up the hill beside the church. Built long and low with its back to the hillside, it commanded extensive views over the valley below. Today this unassuming house which, but for the Arundell arms above the porch and on the gutter heads, gives no hint of its past magnificence, is camouflaged by surrounding orchards and trees. In the small garden by the main entrance stands a very fine early

17th- and 18th-century additions at the rear of Lanherne.

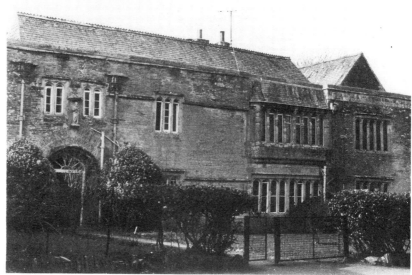

Lanherne, St Mawgan. The Elizabethan entrance front.

Christian cross. It is the fitting attribute of a house whose inmates so steadfastly followed the dictates of their faith and suffered for it. The nuns cherish its presence.

TRERICE, an important and well-preserved Elizabethan manor house about three miles south-east of Newquay, was for four hundred years

Lanherne. The entrance porch with rainwater heads bearing the Arundell coat of arms.

the home of another family of Arundells. Closely related in the Middle Ages to the Arundells of Lanherne, the Trerice Arundells share the same coat of arms. Both bear six swallows and it is from the French *hirondelle* that the name derives.

Ralph Arundell married the heiress of Trerice in the fourteenth century and his descendant John, who died defending St Michael's Mount in 1471, was knighted for his services to the King. The family quickly became one of the most prominent in Cornwall. The 4th Sir John was the builder of Trerice. He married Katherine Coswarth, who brought with her substantial property in the adjoining parish. The actual manor house of Coswarth, however, was entailed on her great-uncle who had set about rebuilding it. Spurred on by his example, or anxious to outdo him, she and Sir John built the present Trerice at about the same time. Work was completed in 1573, a year before Sir John was appointed Sheriff of Cornwall. Richard Carew married their eldest daughter Juliana and often stayed at Trerice. He speaks so

Trerice, near Newquay. The east front, completed in 1573.

warmly of his father-in-law in his *Survey of Cornwall* that each visit must have been a pleasure. He describes the new buildings as 'costly and commodious', as they certainly were.

A small entrance court, at one time laid out with a formal lawn on either side of a central pathway and herbaceous borders along the walls, forms the approach to the house. Like the east front of Prideaux Place, the east façade of Trerice is E-shaped with three projecting bays, but a quite different visual effect has been achieved. It is built of a local limestone that has turned grey with time but is in fact of a yellowish hue. The curved gables are Flemish-inspired and similar to those at Montacute House in Somerset. All the windows of the façade are mullioned and transomed, the magnificent twenty-four-light window of the hall occupying the whole of the space between the south gable and the porch.

Following the traditional mediæval English plan, a screens passage runs from front to back of the house, with the kitchen and service quarters on one side and the principal rooms on the other. The two-storeyed hall has a plaster ceiling decorated with oak leaves and scroll work. Strapwork divides the flat surface into an intricate pattern and curves downwards at intervals to form attractive globular pendants which catch the light streaming in from the enormous east window. The initials on the ceiling, JA, KA, and MA, are those of Sir John and Lady Arundell and his sister Margaret. Above the screens passage is a small minstrel's gallery off a corridor that runs across the centre of the

Trerice. The hall.

house. A row of small arched openings set at cornice level enabled the musicians to look down into the hall below. There probably would have been a carved screen below the gallery originally, as at Prideaux Place, as well as wainscoting. This was removed some time ago and the walls are now plain white to match the ceiling.

The large drawing-room upstairs has always been the main sitting-room at Trerice. It was the solar of the Tudor house, but Sir John made it grander by using the attic space above to make room for the plastered barrel-vaulted ceiling we see today. It is covered with ornamental strapwork and above the frieze at one end of the room are the arms of Henry Fitz Alan, 24th Earl of Arundel and Knight of the Garter, who married Sir John's daughter Mary. The Trerice Arundells were pleased with this marriage and wanted it recorded for posterity. The huge plaster overmantel with more arms quartering other Arundell marriages is also contemporary and dated 1573. Almost identical plasterwork exists at Collacombe Barton near Tavistock in Devon.*

During the eighteenth century Trerice was let to tenants, but when Sir Thomas Dyke-Acland came into the property in 1802 he used it as his base whenever he was in Cornwall. The Aclands sold the estate in 1915 and it was acquired by the Cornwall County Council immediately after the First World War. The 500-acre manor farm was divided into lots which were sold off separately, leaving the house with twenty acres. This was finally bought by the National Trust in 1953 and has been maintained superbly ever since.

The formal Elizabethan gardens round the house have long since disappeared and the original designs are lost. Instead, the Trust has planted fruit trees on the south side, arranged in the quincunx pattern in favour during the seventeenth century, whereby each tree is aligned with the others from wherever it is viewed, and the herbaceous borders are memorable.

LANHYDROCK, another National Trust property, stands in its own magnificent park on the far western slope of the valley of the river Fowey as it turns southwards towards Lostwithiel and the sea. One of the largest houses in Cornwall, it was built in the 1630s by Sir Richard Robartes, a wealthy tin and wool merchant from Truro. Originally a monastic estate belonging to the Augustinian priory of St Petroc, it was obtained by the Glynn family at the time of the Dissolution. A Glynn daughter brought it in marriage to a James Lyttleton and his sister brought it to her husband Thomas Trenance. Their son, Lyttleton Trenance, sold Lanhydrock to the Robartes family in 1620. All these owners lived in a modest monastic barton which probably stood to the north-west of the little churchyard of St Hydroc church immediately behind the present house. But the Robartes made a fortune in the tin trade and Sir Richard Robartes, who paid James I's favourite, the Duke of Buckingham, £10,000 in return for having him made a peer, set about building a grand new house. He did not live to see it finished and his son John, the 2nd Lord Robartes, completed the task.★

The architecture of Lanhydrock, essentially Tudor in appearance with its crenellated parapet, mullioned and transomed windows and pinnacles, shows what scant attention the 2nd Lord Robartes paid to new trends. After all, as leader of the Cornish Parliamentarians in the House of Lords and therefore often in London, he must have seen Inigo Jones's Banqueting House (1619-22) a few hundred yards away in Whitehall, but like most of his comtemporaries failed to appreciate its grace and originality. Robartes preferred a house in an English rather than Italian mould. Built of the local pale grey granite, which does not lend itself to elaborate detailing, the exterior of Lanhydrock has only the simplest embellishments. The hood-moulds of its fine mullioned and transomed windows are joined right the way round the façade by a continuous string course whose horizontality is reinforced by the line of the parapet above. It is interesting to note that the north façade of Godolphin House in Kerrier shares these two features and is thought to be of mid-seventeenth-century date.

The mansion originally consisted of four wings built round a quadrangle, the east wing enjoying views of the gatehouse and the great avenue of sycamores beyond. This whole wing was pulled down towards the end of the eighteenth century, leaving the three sides of the square we see today. At the same time the walls which flanked the

Lanhydrock, near Bodmin. Rebuilt to the original 1630 design after a fire in 1881.

Lanhydrock. The mid-17th century gatehouse and the formal terraced garden laid out in the mid-19th century.

gatehouse and enclosed a forecourt were removed. The gatehouse, with its ball-topped obelisks and Renaissance-inspired arches and columns, now stands in solitary splendour between the park and the formal terraces round the house. The park is a largely seventeenth-century creation but the terraces were laid out by Thomas James Agar, later created Baron Robartes of Lanhydrock (the previous title having lapsed in 1764) in the mid-nineteenth century. Stately rows of cone-shaped Irish yews flank the terrace immediately in front of the house, while the slightly lower terraces are enclosed either by a castellated stone wall or by low box hedges. Within these confines geometrically patterned flowerbeds planted mainly with roses break up the smooth grass parterres. The magnificent bronze urns, decorated in the Baroque manner with cherubs, heavy swags of fruit, animal masks, and vines in high relief, are thought to be the work of Louis XIV's goldsmith Louis Ballin. The fashionable London architect George Gilbert Scott advised Lord Robartes on the lay-out of these elaborate terraces and also enlarged the house, adding a coach-house and stables to the south-east.

On a stormy day in April 1881 the kitchen chimney caught fire. The flames quickly spread and gutted the entire house apart from the north wing and the entrance porch. The Robartes were rescued but Lady Robartes died a few days afterwards from shock. Bereft at the loss of his wife and his house, Lord Robartes fell into a decline and died the following year. Happily, their son Thomas Charles rebuilt Lanhydrock in the original manner.

Inside, to comply with the Victorian demand for comfort, an underground boiler house was installed to heat water (a reservoir filled from a nearby source was built on top of the hill behind the house and provided a constant supply) for household use. The boiler also fed hot water to a network of radiators, some of which are superb specimens with oak cases and marble tops. Electric lighting was put in, and the lay-out of the interior radically altered, particularly with regard to the servants' quarters. As Mark Girouard points out in his *Life in the English Country House*, the Victorians did not necessarily employ more servants than before, but their accommodation changed 'to make them more moral and more efficient'. In the name of morality men and women were kept apart at all times, except when under supervision. Nothing was left to chance. The attic bedrooms for the male and female servants at Lanhydrock are reached by separate staircases. The servants were reminded of their moral duty when they assembled with the family for prayers in the prayer room. There was little opportunity for even a brief encounter with the opposite sex during working hours. The butler's premises were in one part of the house, the housekeeper's in another. The footmen and housemaids worked, as it were, in different domains and would only meet at mealtimes in the servants' hall.

No expense was spared in the building of new kitchen quarters in the south-west corner of the house. A scullery, a dairy, three larders for meat, fish, and groceries, and a bakehouse are grouped round an enormous kitchen. Clerestory windows high up in the roof (they are opened by ingenious mechanisms on the end wall) as well as large windows below ensure not only a light kitchen but a fresh-smelling one. As a further precaution, fumes from the huge coal fire could escape through metal grids in the wall above. Food was kept hot in a serving-room next to the dining-room. The dishes were placed in the massive iron cupboard heated by hot water pipes which is still there today.

The Victorians had a passion for segregation. To judge from the number of backstairs and back corridors at Lanhydrock, the staff could go about their business without bumping into members of the family. They were not to be seen and not to be heard, rather like the children, who, of course, were banished to the nursery wing. Even Lord and Lady Robartes had their own demarcation lines. The oak-panelled billiard room and smoking room on the ground floor were a strictly male preserve where the men would gather after dinner. They are hung predictably with old photographs of school and college teams and with engravings of Lord's cricket ground. On the first floor, well away from the men's social gatherings, is Lady Robartes' suite of rooms, comfortably furnished with buttonback chairs, William Morris wallpaper and much cheerful bric à brac. If her boudoir symbolises Victorian cosiness and clutter, the large drawing-room nearby, whose furniture is in fact largely Georgian, represents Victorian formality.

A small flight of steps at the end of the drawing-room leads into the enormous Jacobean gallery that survived the fire. It is 116 feet long and runs the length of the north wing. The plasterwork of the magnificent

Lanhydrock. The Jacobean long gallery.

barrel-vaulted ceiling is well worth a close look. The main panels, twenty-four in all, are star-shaped and depict in lively fashion various Old Testament subjects. Smaller panels portray a wonderful array of birds and beasts, some real, some imaginary. But when viewed as a whole, it is the amazingly intricate pattern of strapwork that catches the eye. Whoever the plasterers may have been (the National Trust attributes this ceiling and the one at Prideaux Place 'almost certainly' to a Devonshire family of plasterers, the Abbots of Frithelstock near Bideford), they had a superb sense of design as well as imagination.

The Higher Garden, behind the house and little church, makes a good starting-point for a tour of the garden and the vast park, which was laid out in 1860 at the same time as the terraces, by Thomas James Agar. Many of the original flowerbeds on the lower slopes have disappeared and the flowering shrubs (the usual mixture of magnolias, azaleas and rhododendrons) are largely twentieth-century additions. A curious assortment of buildings provides support for many of the plants. Flowering quinces, firethorn, and a trumpet vine grow up the wall of a picturesque thatched cottage. Nearby is a small, partly sunken edifice whose quaint gables end in tight twirls, like a roll of granite carpet, marking the site of the Holy Well used by the monks of St Petroc long, long ago.

A wide, winding path leads from the Higher Garden up the hill behind the house. Clusters of many coloured rhododendrons grow on the grassy slopes, together with specimen trees, and a pink magnolia forms an archway over the path itself. Graceful beech trees crown the summit, and while enjoying their shade you can look down, past the kaleidoscopic colours of the Higher Garden, to the solitary gatehouse and the great avenue of sycamores beyond. These were planted in 1648 by the 2nd Lord Robartes, and in the nineteenth century a row of beeches was planted next to them, so that the approach to the house would still be lined with trees when the sycamores died. A small gateway marks the entrance to the Great Wood, which offers a number of different routes back to the house. Giant-leaved Himalayan rhododendrons and plantations of azaleas greet the eye, as well as thousands of bluebells.

Thomas Charles Robartes, who rebuilt Lanhydrock after the fire, succeeded to the Viscountcy of Clifden in 1899. His eldest son, Tommy Agar-Robartes, was a prominent Cornish Liberal and much loved in the county. As an officer in the Guards he died gallantly on the Western Front in the First World War, attempting to save the life of a wounded comrade, for which action he was recommended for a posthumous VC. There is a moving memorial to him in Truro Cathedral. The second son, Francis Gerald, became the 7th Viscount. Francis never married and neither did his two sisters. They lived together at Lanhydrock for over thirty years and gave the house and park to the National Trust in 1953. It is to be hoped that the monks of St Petroc would have approved of the pleasure Lanhydrock gives to so many today.

7.

Powder

The whole castle beginneth to mourne, and to wringe out
harde stones for teares, that shee that was imbraced, visited
and delighted with greate princes, is now desolate, forsaken,
and forlorne.

John Norden, *Description of Cornwall*, 1610

RESTORMEL CASTLE, so poignantly described by Norden, lies in the
north-east corner of Powder Hundred about a mile to the north of the
old town of Lostwithiel. In the Middle Ages the river Fowey was still
navigable at this point, and Lostwithiel was an important trading
centre. It is thought that the Norman lord, Baldwin Fitz Turstin, who
held many manors for Count Robert of Mortain, bridged the river
Fowey in about 1100 and built a fort high above it to protect the new
crossing. Turstin's domain was worth guarding, for the weekly
markets at Lostwithiel meant that handsome sums of money, tolls for
the right to trade and rents from each stall holder, went into the owner's
pocket. Richard, Earl of Cornwall, naturally wished to gain control of
such a valuable asset. He had already secured Launceston and Tintagel
and in 1270 he persuaded Isolda de Cardinan to sell him her Restormel
estates and the borough of Lostwithiel.

Restormel was the favourite residence of Richard's son Edmund, the
next Earl of Cornwall, and it was he who decided to make Lostwithiel
his seat of government in the county. Edmund constructed a splendid

Restormel Castle (late 12th century), Lostwithiel.

Restormel Castle. The great hall (late 13th century)

range of buildings for his new centre of administration. There was a hall of the exchequer for the payment of dues; a coinage hall where a coign or corner was cut from each tin lode for assay; a stannary hall for the evaluation of tin and a stannary court where the disputes and debts of the tinners were settled. Fragments of these old buildings can still be seen in some of the houses in the town.

In the Middle Ages Restormel Castle stood in a vast deer park. Nothing remains of Baldwin Fitz Turstin's castle except some early masonry at the base of the gate tower. The circular wall of the shell keep was erected about a hundred years later, at the end of the twelfth century. Built of local slate with dressings of white Pentewan stone, it is well protected by a deep flat-bottomed ditch below and by a battlemented parapet round the eight-foot-wide wall-walk above. This was the castle which Richard bought from Isolda de Cardinan and which his son Edmund enlarged and refined.

Inside the keep, Edmund erected a series of rooms for his personal use. The partition walls survive today, so that it is possible to visualise those luxurious new living quarters. To the right of the gate is the kitchen, the only room which rises to the full height of the castle. The colossal fireplace built into the curtain wall, as well as the adjoining buttery and pantry, are evidence of the Earl's hospitality. All the other rooms at ground floor level served as store rooms and have low ceilings. The principal rooms – the great hall, the solar, the ante-chapel and two bedchambers, as one proceeds in an anti-clockwise direction – were built above these storerooms, away from the courtyard bustle and with easy access to the wall-walk. Large windows cut through the curtain wall gave each room a wide view over the hills and woods of the surrounding countryside; log fires would have provided warmth and a flickering light as darkness fell.

The bailey or outer court on the west flank of the hill comprised servants' quarters, stables, offices, storerooms, and a chapel. From the higher ground above the bailey, water from various springs was channelled through a leaden conduit into the moat and keep. A deep well in the courtyard inside the keep, the original water supply of the castle, was retained for use in an emergency, should the conduit be damaged during an attack.

After Edmund's death in 1299 Restormel was virtually unused. The Black Prince visited the castle in 1354 and 1365, in search, one suspects, of money rather than of pleasure. He wished to collect his feudal dues and to raise money and troops before setting off for Bordeaux, where he maintained the most glittering court in Europe. Like the other strongholds in his Duchy, Restormel deteriorated; although briefly held by Lord Essex for Parliament during the Civil War it was never used again.

BOUND by Falmouth estuary to the west and by Fowey estuary to the east, Powder Hundred's coast is well protected by **St Mawes Castle** and **St Catherine's Castle**, which guard the approaches to these two great rivers. Both were built by Henry VIII in the late 1530s and 1540s as part of his defence of the realm. ★

Earlier in his reign Henry had built small blockhouses at St Mawes to stop enemy and pirate ships from entering the river Fal, but he now ordered a fortress to be built close to the water. Thomas Treffry, Clerk of Works to the King, was in charge of the operation, and a military architect from Moravia, Stefan von Haschenperg, is known to have been employed by Henry in the 1530s. His ideas were brought into play at St Mawes. Work began in 1540. The design of the castle was unlike anything the English had seen before. The invention of cannon and gunpowder had rendered tall keeps and gatehouses obsolete. Nor was St Mawes ever intended as a nobleman's residence, as most mediæval castles were. Three low, semi-circular bastions surround the central tower, which served as a garrison for as many as a hundred men. The mess-room was on the ground floor, and the kitchen in the basement. Large cannon were placed in the courtyards of the bastions, and light guns in the embrasures on the rampart walls. There are eight further recesses for gunners in the upper room of the tower. Entrance to the castle was via a drawbridge on the landward side. Latin inscriptions singing the King's praises appear on the walls of the keep, as well as lively gargoyles and an elaborate royal coat of arms.

Michael Vyvyan of Trelowarren became governor of St Mawes Castle in 1544. This was an onerous task since he was required to

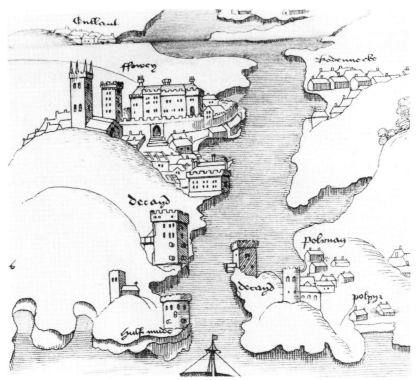

Fowey Haven. Part of a chart drawn in the reign og Henry VIII showing St Catherine's Castle at the harbour entrance, the toll or south gate of Fowey and, next to the church, the battlemented walls of Place.

maintain the garrison at his own expense. Nearly a hundred years later this unrewarding position proved too much for his descendant Sir Francis Vyvyan, governor in the early seventeenth century. He was dismissed in 1632, and it was left to his successor, Hannibal Bonython, to rescue St Mawes from dire poverty. Fourteen years later he surrendered the castle to Fairfax's Parliamentary force without a fight. He can hardly be blamed, for the garrison was poorly equipped for a siege without arms or sufficient ammunition. This inglorious action was the castle's first and last.

Fowey, with its maze of tiny streets rising steeply from the huge natural harbour, was always prone to attack by virtue of its importance as a port and as a haven for shipping. During the fifteenth century the town thrived on privateering and piracy. The renowned 'Fowey gallants' used to attack French ships off the Breton coast and strip them of their cargoes. Reprisals were inevitable: in 1457 French pirate ships sailed into the harbour under cover of darkness. The inhabitants rallied to the town's defence under the direction of Elizabeth Treffry of Place, whose house afforded them a degree of safety as the marauding French set fire to the houses below. Having learned a bitter lesson, the men of Fowey constructed two blockhouses on either side of the mouth of the river and suspended a heavy chain between the two as a way of at least damaging enemy vessels. A century later Henry VIII erected St Catherine's Castle on the south side of the harbour as a further precaution against intruders.

Of simpler construction than St Mawes Castle, it consists of a single round tower, which housed six cannon. It is best seen from the other side of the water at Polruan.

PLACE, still the seat of the ancient Treffry family whose history is inextricably linked with that of Fowey, is scarcely visible from the town or the waterfront. Its battlemented walls appear to circumscribe the town rather than a particular house. The Treffry family was well established there by the late Middle Ages.★

John Leland, Henry VIII's antiquarian and chaplain, stayed at Place in 1538, perhaps in order to keep an eye on Thomas Treffry as he oversaw the construction of the coastline defences already mentioned. He commented that an earlier Thomas Treffry, husband of the redoubtable Elizabeth who had repulsed the French, 'builded a righte fair and stronge embatelid tower in his house and embatteling all the walles of the house in a manner made it a castelle'. This description matches a drawing of Place, made on a contemporary map, which shows a house with a tower surrounded by a high battlemented wall.★★

Joseph Thomas Austen, who inherited the property through his mother in 1786, rebuilt Place between 1813 and 1845. Romantic on the one hand, traditionalist on the other, he chose the Regency Gothic style as a way of remaining faithful to the mediæval Gothic of the original house, much of which remains. The Tudor courtyard, with its handsome oriel window, is still the centre of Place, although the rooms around it were rearranged. The hall became the dining-room, the solar became the library. Very sensibly, Joseph Thomas Treffry – he assumed the name when High Sheriff in 1838 – restored the magnificent oak-beamed ceiling of the hall and preserved the mid-eighteenth century ceiling in the library. The beautiful rococo plasterwork of the latter was executed by Mr Heyden and Mr Lorington who had been working at Carclew for the Truro mining magnate William Lemon. The Regency Gothic bay window in the library, with its decorative tracery, was inspired by the Tudor bay window of the morning room. In such thoughtful ways were past and present linked. Lastly, Mr Treffry added the granite towers, thus maintaining the fortified appearance of old Place. The entrance hall was built entirely of porphyry from his own quarries.

The gardens at Place were redesigned in about 1900 by the landscape gardener T. H. Mawson. Captivated by the natural beauty of the site, Mawson created a series of terraces and walks running parallel to, but high above, the river. These flanked one side of the long carriage drive. Overgrown long since, it is the present owner's wish to recreate them

Place, Fowey. 19th-century engraving of the south front with the library and drawing-room bow.

in part.

If the remodelling of Place allowed the remarkably intelligent and re-sourceful Mr J. T. Treffry to indulge his fancy, his other activities reveal an intensely practical side to his character. With great energy and enterprise, he became the largest employer of labour in the West Country, owning mines, granite quarries and railways. He also built Par harbour and the present main road into Fowey. He was deeply concerned with the early development of railways in Cornwall and constructed the line from Par to Newquay. As the first Chairman of the Cornwall Railway, he was closely associated with the engineer I. K. Brunel in the project for building the great railway bridge across the Tamar, although this was not completed until after his death. Perhaps Joseph Thomas Treffry is best remembered for the construction of the **Treffry Viaduct** across the deep Luxulyan valley to the north-west of Fowey, its purpose being to carry water and a light railway, now disused. This colossal granite structure, nearly 700 feet long and 700 feet high with 10 gigantic arches, is a tribute to its creator.

An enchanting, exceptionally well preserved late mediæval farmhouse called **Methrose** lies about a mile to the south-west of the viaduct. Situated in a dip of land at the end of a very narrow lane and surrounded by a high granite wall partly covered with ivy, Methrose is easy to miss. Its unobtrusive, almost ramshackle appearance is part of its charm. A four-centred arch opens onto a minute courtyard bright with honeysuckle and geraniums in summer. The farmhouse occupies two sides of the courtyard. On the right is the old Tudor house with its low walls, uneven drooping roof, and tiny oriel window. The hall, which is still the principal living-room, rises the full height of the house; the service end is divided into two storeys with the solar above. Facing north across the courtyard is a parlour wing, thought by the Cornish historian Charles Henderson to have been built between 1622 and 1649 by Nicholas Kendall. This striking, newer building, with its symmetrically disposed windows, straight drip course, and tall gable end chimneys, lacks the character of the older house, which has assimilated the changes round it with equanimity. Methrose lives on because it is still farmed, just as it was centuries ago when a comparatively well-to-do family settled there and became the Methroses of Methrose.

ONLY the scantest remains of another late Tudor house can be seen at nearby **Prideaux**, to the west of Luxulyan valley. Its fate was sealed in 1808, when a member of the prosperous Rashleigh family built a new house a few hundred yards away with fine views towards St Austell bay. This handsome rectangular edifice with large, bright rooms rendered the old manor obsolete and, like an untended plant, it soon withered and died.

COMPARED to the Treffrys, the Rashleighs were newcomers when, as ambitious Devon merchants, they came to trade in Fowey in the sixteenth century. In 1545 Philip Rashleigh bought the manor of **Trenant**, just a few miles to the west of Fowey. It originally belonged to the priory of Tywardreath, but after the Dissolution monastic lands were sold off in order to raise money for Henry VIII's disastrous wars. Philip was quick to seize this golden opportunity to acquire land, and his son John added to the estates by buying Bodmin Priory in 1567. Although a furnished chamber was kept clean and aired for his use, he never made the priory a permanent home. It was divided up into separate tenements and soon fell into a state of disrepair.

JOHN RASHLEIGH decided early in his career to make Fowey his base, and his town house, now the **Ship Hotel** at the bottom of Lostwithiel Street, looked on to what is now called Trafalgar Square. John and his wife Alice renovated the house in 1570 and the oak-panelled room with their names carved on the mantel has been preserved. Caryatids and other Renaissance motifs show local awareness of Italianate ideas. A doorway that still exists in the upper storey used to lead into a room in the arch across Lostwithiel Street. This was the toll gate or south gate of the town, pulled down in 1876. The existing front of the Ship Inn is late nineteenth century.

IN about 1600 John built himself a fine new mansion on the land he had acquired between Fowey and Gribbin Head. **Menabilly**, now the home of the Rashleigh family, was entirely rebuilt in the early eighteenth century and substantially altered in the 1820s when the grounds were landscaped. Sheltered by woodland and stretching down to a secluded bay, these were planted with sub-tropical plants and rare shrubs.

For many years Menabilly was the home of the writer Dame Daphne du Maurier and her husband, the late General Sir Frederick Browning. Its secluded atmosphere must have contributed to the character of Manderley in her celebrated novel, *Rebecca*; and Menabilly is at the centre of her less well-known but outstanding novel of the Civil War, *The King's General*.

TREWITHEN, mid-way between St Austell and Truro in the parish of Probus, must take pride of place as the finest eighteenth-century house in Powder. Building began not long after Philip Hawkins, a wealthy attorney's son, bought the estate in 1715. The Cornish historian Thomas Tonkin noted, in the *Parochial History* that he began in 1702, that Trewithen Barton had previously belonged to the Williams family but that Courtenay Williams had foolishly squandered away 'a pretty estate and a good fortune too', and sold the barton to Philip Hawkins. Philip thereupon 'very much improved this seat, new built a great part of the house, made good gardens etc.'*

The north or entrance front of Trewithen, with its recessed centre, is built of bricks made from blue clay dug on the estate. The porosity of the bricks meant that after a period of weathering the brickwork had to

Trewithen. The north front, designed by Thomas Edwards for Philip Hawkins c. 1715.

be sealed. In 1948 the north front was therefore re-rendered with a special concrete mixture (the sand came from Hayle Towans, owned by the Hawkins in the eighteenth century), which happened to have the same ingredients as Pentewan granite. As a result, the concrete facing blends perfectly with the other three sides of the house which are built of solid Pentewan stone. This stone is a soft silvery grey on a dry day, but when the air is laden with moisture it has a definite pinkish hue. Sombre granite keystones above each sash window are the only decorative feature of the otherwise plain north front, but its severity is subtly tempered by the presence of two detached wings of pinky red bricks surmounted by matching cupolas, one of which contains the oldest single-handed clock in Cornwall. These attractive buildings stand at right angles to the house and frame the forecourt in a firm but casual way. The east wing was originally a stable block, while the west wing contained a laundry and bakehouse, the estate office, and a court room where the annual rent dinner was held.

Philip Hawkins had no children and his cousin Thomas succeeded to the estate in 1738. The south or garden front was probably built in his time and is thought to be the work of the architect Sir Robert Taylor, possibly following Edwards's design. It is grander than its north-facing counterpart.

The cental block projects rather than recedes in a gesture of assertion. Its elegance is due to the strong but simple stone cornice and the delicately carved consoles beneath the window frames. The central window on the ground floor is further stressed by the addition of an imposing entablature supported by similar but larger, consoles.

A magnificent dining-room fills the five central bays of the south

Trewithen. The south front.

Trewithen, Probus. Engraving from Borlase's Natural History of Cornwall (1758) showing the rows of beech trees planted by Thomas Hawkins on either side of the south front in the mid-18th century.

front, its rococo stucco decoration indicating a mid–eighteenth century date. Graceful garlands and floral pendants beneath candle brackets break up the wall space, painted a soft Wedgwood green, while the length of the room is relieved by triple arches at either end which form a vaulted arcade. Each arch is supported by Ionic columns matched by Ionic pilasters on the wall which offset the various portraits. Two of the finest, by Sir Joshua Reynolds, are of Dr Zachariah Mudge and his wife. Mrs Mudge, it will be remembered, severely scolded Dr Johnson for drinking his seventeenth cup of tea. The Mudges became associated with Trewithen by marriage, and another Reynolds portrait of Kitty Mudge, as the Market Girl, hangs in the drawing-room. There are also four fine portraits of the Hawkins family by Allan Ramsay, an unusual choice of artist in Cornwall.

Thomas Hawkins married Ann Heywood, the daughter of a well-to-do London cloth merchant. Her sophisticated taste and fashion consciousness are reflected in the elegant interior of Trewithen. As a young bridge she would have supervised the decoration of the dining-room and drawing-room, the one in the rococo manner, the other with Chinese Chippendale fretwork over the handsome doorways. The craftsmanship is superb, whatever the detailing. Ann was a perfectionist for whom nothing but the best would suffice.

If Ann's taste is manifest inside Trewithen, Thomas's influence lives on outside. He was responsible for the two detached wings and for landscaping the garden. An engraving in Borlase's *Natural History of Cornwall* shows the garden as it was in his time. Rows of beech hedges flank each side of the south front to create a formal vista and to protect the house from the prevailing wind. Avenues radiate from the house to the east and to the north, with a landscaped park beyond. But Thomas did not live to see his trees mature. Having set an example to the wary villagers by having himself vaccinated against smallpox, he died of the disease in 1766, at the age of only forty-two.

His eldest son Christopher, created a baronet in 1799, inherited Trewithen and greatly increased the family's fortunes.★

He died a bachelor in 1829 and his nephew Christopher Henry Thomas Hawkins succeeded him. He was the last Hawkins to inhabit Trewithen, for on his death in 1903 the estate passed, through his sister Mary Ann, to the Johnstone family. George H. Johnstone, grandfather of the present owner, was largely responsible for the magnificent gardens we see today. When he inherited the estate in 1904, Thomas Hawkins' beech hedges were mature trees. Three hundred of these were felled by government order during the First World War, which meant that openings were made in the wood on each side of the south front. These empty spaces cried out to be filled, and George Johnstone was convinced that the more tender species of rhododendron, camellia and magnolia would flourish in such a sheltered environment. Trewithen today leaves us in no doubt that he was right. It was the beginning of a golden age of planting.

The setting for a vast carpet of lawn, over two hundred metres long, which rolls away from the south front towards a woodland shrubbery is George Johnstone's greatest achievement. This immense glade is lined on both sides by a dazzling range of trees and shrubs, many of which were raised from seed collected in the wild by the great pioneer plant collectors at the turn of the century. Magnolias came from E. H. Wilson's expedition to China in 1899, camellias from George Forrest's Himalayan expedition, and rhododendrons from Kingdon Ward's explorations in Assam and Burma. Among the most beautiful are the white *Magnolia mollicomata*, now a tall tree; *Magnolia campbellii* with deeper pink flowers; the famous brilliant yellow *Rhododendron macabeanum*, collected in Manipur by Kingdon Ward, and the white or pinkish *Camellia saluenensis*, brought back by Forrest.

George Johnstone not only introduced these exotic plants to Trewithen but raised many hybrids himself, some of which are named, endearingly, after those close to him. *R.* 'Alison Johnstone' is named after his wife, *R.* 'Jack Skilton' after his head gardener and *C.* 'Elizabeth Johnstone' after one of his daughters. He also bred an astonishing variety of daffodils. The yellows range from palest primrose to Van Gogh gold, but there are whity pinks, creamy whites and even tinges of blue to be seen at close range.

This passionate plantsman loved to experiment, as a painter does, with form and colour. Like any artist concerned with visual beauty, he aspired to create a unified, balanced composition. Great care and thought have gone into the design of the gardens. Each area is different, yet related to the whole. To the east of the great glade is a path leading to an avenue of sycamores which commands a fine view across the park. Beside the path is a row of specimens of two very primitive trees, *Metasequoia glyptostroboides* or Dawn Redwood and *Ginkgo biloba* or Maidenhair tree. A male specimen of the latter was planted at Kew in

1762 and is there to this day. The branch of a female grafted on to it in 1911 produced seeds eight years later. The Trewithen ginkgos were planted soon afterwards.

To the west of the glade is the camellia walk. Here are magnificent examples of *C. saluenensis* and *C. reticulata forma simplex* grown from Chinese seed, successful clones such as *C.* 'Trewithen Pink', *C.* 'Trewithen Salmon', *C.* 'Elizabeth Johnstone' and many forms of *C. x williamsii*, to be mentioned later in connection with Caerhays. Dwarf rhododendrons planted alongside the camellias burst into bloom in May. The camellia walk leads to the Cock Pit, a deep dell filled with rhododendrons and primulas as well as tree ferns from New Zealand, a Japanese maple and large pink *Magnolia sprengerii*.

In contrast to this exotic underworld full of riotous colour, the small walled garden that fills the angle between house and stable block on the west side is of a quieter beauty. Formality and intimacy come as a relief after the riotous planting 'in the wild'. A slightly sunken ornamental pond at one end is complemented at the other end by a raised terrace complete with summerhouse and rose trellis. More roses fill the flowerbeds which break up the rectangular grass parterre in the centre. This is surrounded by a path of granite setts from the old Camborne-Redruth tramway, lovingly laid out by George Johnstone himself. Quintessentially English borders packed with pinks, lupins, michaelmas daisies, and delphiniums in high summer flank the long sides of the rectangle and make a contrast to shrubs.

In December 1979 a terrible storm caused havoc at Trewithen. It was a terrifying night. Eighty-eight trees crashed like ninepins as the wind tore through the beech wood behind the walled garden. The destruction was overwhelming, the task of clearing the trees and saving the battered shrubs daunting. Mr and Mrs A. M. J. Galsworthy made a brave decision. They felled the remaining trees, which were anyway nearing the end of their lives, and planted new shrubs in their place. West of this section they have created an arboretum to provide the necessary shelter in years to come. Pines and eucalyptus from all over the world have been collected and planting still continues.

THREE great nineteenth-century mansions in Powder - Tregothnan, Trelissick, and Caerhays - are the visible expression of the vast fortunes made by managers and landowners alike from the Cornish tin and copper mines.★

Foremost among these families were the Boscawens of **Tregothnan**. Their huge house stands on a hill above the east bank of Tresillian river, a tidal creek that joins the Truro river and then flows into the Fal. Tregothnan means 'the house at the head of the valley' and is the seat of Lord Falmouth today.

The Boscawens originally lived at Boscawen Ros in the parish of St

Buryan, near Land's End, but when John de Boscawen married the heiress Joan de Tregothnan in 1335, he moved to Tregothnan where his descendants have continued to live. The original Tudor house, a two-storied building with a battlemented tower and arched doorway beneath, lay to the north-west of the present terrace where a number of ilexes grow. All that remains is the old doorway which stands at the entrance to the kitchen garden.

In the 1650s Hugh Boscawen built a fine new house which his cousin Celia Fiennes (Hugh's first wife was Margaret Fiennes, daughter of the Earl of Lincoln) describes in her journal of 1698 when she travelled to Land's End on horseback. How glad she must have been to be 'civilly entertained' by her relatives after riding in wet weather along rough roads. 'The house', she noted, 'is built of white stone like the rough coarse marble [Truro porphyry] and covered with slate; they use much lime in their cement which makes both walls and cover look very white.' It was a rectangular block with a hipped roof, typical of a type of house built during the Commonwealth and which Sir John Summerson refers to as 'artisan mannerist' in style.

The entrance, [Celia goes on] is up a few stone steps into a large high hall and so to a passage that leads foreright up a good staircase; on the right side is a large common little parlour for constant eating in, from whence goes a little room for smoaking that has a back way into the kitchen and on the left hand is a great parlour and drawing-roome wainscoted all very well, but plaine.

Hugh Boscawen's nephew inherited Tregothnan and was created Viscount Falmouth in 1720 for his services to George I as Controller of the Household. His son John married Joan Blanchland, who brought with her a large estate on the opposite side of the Fal. Soon afterwards rich lodes of copper were found on this land and successfully mined, thereby greatly increasing the Boscawen fortune. The 4th Viscount Falmouth, created 1st Earl of Falmouth at George IV's coronation, employed William Wilkins, architect of the National Gallery in Trafalgar Square, to enlarge Tregothnan in the 1820s. Wilkins turned the sober seventeenth-century house into a picturesque 'Tudor' castle. An imposing three-sided 'gatehouse' acts as a huge covered entrance porch in the middle of the north front, its Gothic windows and arches an effective contrast to the square mullioned and transomed windows on either side. The curiously shaped pinnacles on top of the angle turrets are replicas of the pinnacles of East Barsham Manor, a late tudor house in Norfolk. In 1845, another architect called Vulliamy enlarged the south front for the 2nd Earl, adding a gabled wing and towers at the west end.

When Celia Fiennes visited Tregothnan the garden was hardly established. It had 'gravel walks round and across', she recorded, 'but the squares are full of gooseberry and shrub trees and look more like a

Tregothnan, near Truro. 19th-century engraving of the entrance front, designed by William Wilkins for the 1st Earl of Falmouth c. 1815.

kitchen garden as Lady Mary Boscawen told me, out of which is another garden and orchard which is something like a grove, green walks with rows of fruit trees; it is capable of being a fine place with some charge . . .'

Evelyn Boscawen, the 6th Viscount Falmouth, was largely responsible for the layout of the present gardens which cover an area of about forty acres. In the 1850s and 1860s he and his brother John, the Rector of nearby Lamorran, planted many varieties of *Camellia japonica* near the east front of the house. These were removed to the far end of the garden soon after 1897 where they now flourish at the top of a slope known as Snowdrop Hill. They include 'Lady de Saumarez', 'Wilbankiana', 'Arejishi', 'Hornsby Pink', 'Canon Boscawen', 'Tricolor' and 'Imbricata Alba'. Apart from 'Lady Clare' and 'Nagasaki', no new plantings of the modern camellia varieties were made until 1956 when C. 'Magnoliae flora', 'Salutation', 'J. C. Williams' and 'Donation' were planted. In recent years new varieties have been planted including 'Drama Girl', 'Prince Albert', 'Apple Blossom' and 'Golden Spangles'.

In front of the summer house is a large expanse of grass flanked by vast clumps of blood-red *Rhododendron arboreum*. Well over one hundred years old, these are now almost fifty feet high. Other rhododendrons include a number discovered by Joseph Hooker in 1848 and described in his book *The Rhododendrons of the Sikkim-Himalayas*. Two of the most beautiful are *R. falconeri*, a tree with creamy bell-like flowers marked with purple in the throat found near Darjeeling, and the beautiful yellow *R. campylocarpum* from Sikkim.

Mansion House, Truro. Designed by Thomas Edwards for Thomas Daniell c. 1750.

THOMAS DANIELL began his career as clerk to the Truro mining magnate William Lemon, but later made a fortune in his own right. In the 1750s he employed the architect of Trewithen, Thomas Edwards, to build the **Mansion House** in Truro. His wife's uncle, Ralph Allen of Bath, made them a gift of stone from his own quarries to build the new house. Delivered by barge to Daniell's quay on the river at the bottom of the garden, it launched an immediate vogue for Bath stone, which was subsequently used for houses in nearby Lemon Street.

IN 1800 Thomas Daniell's son, Ralph Allen, acquired **Trelissick,** a modest mid-eighteenth century house designed by Sir Humphrey Davey's grandfather for a captain in the county militia. Ralph Allen's son, Thomas, rebuilt the house in 1825, adding an imposing but austere neo-classical shell to the existing core. He chose as his architect Peter Frederick Robinson, a pupil of Henry Holland who designed Carlton House for the Prince Regent. Horace Walpole praised the 'august simplicity' of Holland's neo-classical style, where surface decoration was cut back to a minimum in order to enhance the beauty and outline of the overall design. It is no surprise to find his pupil Robinson using the Ionic order at Trelissick, for Holland was particularly fond of it and had already given the Duke of York's splendid Dover House in Whitehall a magnificent Ionic portico. In 1827 Robinson published his *Designs for Ornamental Villas*, and design No. 3 is clearly based on Trelissick.

Thomas Daniell implemented his father's plan to lay out carriage

Trelissick, near Truro. Remodelled by P. F. Robinson, a pupil of Henry Holland, in 1825 for the Daniell family.

roads through the woods and to plant trees along the Fal estuary. It was an ambitious project whose legacy we enjoy today, but it helped to cripple Daniell financially. His legendary wealth ran out in 1832 and he was obliged to flee to France to escape his creditors. It was a long exile, for he died in Boulogne in 1866. His neighbour across the Fal, the Earl of Falmouth, bought Trelissick but sold it twelve years later to a John Davies Gilbert. Gilbert's son added a second storey to the single storey wings that had been part of Robinson's original design.

When Carew Davies Gilbert died in 1913, his executors let Trelissick to Mr L. D. Cunliffe, one of the governors of the Bank of England; in 1920 he bought the house and part of the estate. He and a French architect, M. Joubert, planned the present solarium. His step-daughter Mrs Ida Copeland inherited the property in 1937 and she and her husband Ronald (whose family firm was W. T. Copeland & Sons Ltd, manufacturers of Spode China) moved from Staffordshire to Trelissick in 1948. The house is now occupied by their son, and Trelissick gardens have been given to the National Trust.

Standing at the head of a vast and very beautiful natural harbour and facing Falmouth across the water, the site of Trelissick is spectacular. The parkland that stretches down to the water's edge was largely planted in the 1820s, although the Gilbert family continued the process. Like the picturesque buildings in a harbour scene by Claude, the graceful silhouettes of the mature oak, beech, and pine trees frame the vista down to the water and gently lead the eye to the middle distance and beyond to the misty waters of the Carrick Roads and the town of Falmouth.

The Copelands were largely responsible for the wonderful gardens to be seen today. Ronald Copeland was a rhododendron expert and planted many of the more tender varieties such as the white and pale pink hybrids raised from Himalayan and Chinese species and *R. griffithianum* or those from *R. cinnabarinum* with their distinctive blue-green foliage. The claret-coloured 'Royal Flush', pink 'Lady Rosebery'

Trelissick Gardens. View over the river Fal.

and apricot 'Lady Chamberlain' are among the loveliest of the latter category. He also planted a large number of the existing camellias and the very mild climate also made it possible to grow tender South American shrubs. Most eye-catching are the creamy white flowers of *Drimys winteri*, the crimson *Tricuspidaria lanceolata*, and *Abutilon vitafolium*, a climber with lilac blue flowers. *Halesia corallina*, better known as the snowdrop tree, imported into Britain for the first time in 1756 and named after Dr Stephen Hales, a plant physiologist, has white, 'pearl-drop' flowers, which hang in rows beneath the branches and look like air-borne snowdrops waving in the breeze.

CAERHAYS CASTLE overlooks the cliffs of Porthluney Cove, a small bay between Nare Head and Dodman Point. Both castle and site are supremely romantic. The history of Caerhays is full of romance, too, its early owners having been given to chivalrous deeds and fits of passion.

The Trevanions came to Caerhays in the mid fourteenth century when one of their number married Joan Arundell. ★

Caerhays Gardens, Gorran. Designed by John Nash in 1808 for John Trevanion. 19th-century engraving showing the deer park between the castle and the sea.

The property remained in the Trevanion family through direct descent or through marriage until the end of the seventeenth century when it was pulled down.

In 1808 John Trevanion asked the architect John Nash to build a vast romantic castle below the old house with pleasure gardens leading down to the cliffs. Such a project gave both men a chance to indulge their most extravagant fantasies. Today, when a thick mist blows in from the sea and presses round the castellated walls, Caerhays looks like a fairy-tale castle floating in the vaporous air. Towers and turrets, round and square, rise up out of the gloom, their ghostly forms silhouetted against dark, dripping trees.

The interior of the castle with its round drawing-room, double staircase and long gallery, was as grandiose in conception as the exterior. The garden plan was equally extravagant. Formal parterres and terraced walks were laid out within the castle walls, while outside paths led through the woods and deer park to an ornamental lake lying between the castle and the sea. Each scheme was more extravagant than the last and, inevitably, Trevanion's resources ran out. By 1824 much of the estate land as well as the house itself was mortgaged to raise money for the building costs. Trevanion was too much in debt for this to be a solution and he left for Brussels, presumably to escape his creditors. He died there in 1840.

In 1853 the unfinished castle, mortgaged to the hilt, was put up for sale and bought by another rich mining family, the Williams of Scorrier. Lands which had for so long sustained the Trevanions were sold off in lots. The final irony was that much of it proved to be fabulously rich in china clay. The Williams family restored the castle (the original contents were lost in a terrible fire) and remain in possession of it to this day.

The Williamses not only preserved Caerhays but made a significant contribution to English gardening. John Charles Williams came to live at Caerhays in about 1886. 'J.C.', as he was called, was one of the first gardeners to realise the exciting possibilities open to him when plant collectors began bringing back shrubs from China and Asia. The garden is based on the plant importations made by E. H. Wilson between 1900 and 1910 and by George Forrest between 1912 and 1925, the latter sending much seed direct to Caerhays. J.C. made about 250 rhododendron crosses during his lifetime, only three per cent of which have been named. The apricot 'Royal Flush', shell pink 'Veryan Bay', 'Red Admiral', and scarlet 'Humming Bird' are among the most successful hybrids raised at Caerhays. Special rhododendrons include the pink/purple *davidsonianum* introduced by E. H. Wilson in 1908, the large-leaved cream *davidsonianum* introduced by George Forrest in 1913 and the scented whity pink 'Decorum' from Yunnan. J.C. also experimented with camellias, and crossed Forrest's *Camellia saluenensis* with the *Camellia japonica* introduced to England in the mid eighteenth century. This produced the first of a new race of camellias, the x *williamsii* group, the finest of which is the pale pink *C.* 'Donation'. Others include the soft single pink 'J. C. Williams', the deep single pink 'St Ewe' and the double purplish pink 'Caerhays'.

Magnolias from China also flourish in the grounds and look like great water-lily trees. *M. mollicomata* and *M. soulangiana* 'Nigra' were crossed to form 'Caerhays Surprise'; the gentle pink *sprengeri* 'Diva' and *sargentiana* 'Robusta' are the parents of the salmon pink 'Caerhays Belle'. There is also a wide range of evergreen and deciduous azaleas from India, Formosa, Japan and the United States of America.

Under J.C. Caerhays became a gigantic nursery garden for the whole of Cornwall and beyond. His descendants are continuing the planting tradition.

8.

Kerrier

*St Mawes lieth lower and better to annoy shipping, but
Pendennis standeth higher and stronger to defend itself.*
Richard Carew, *The Survey of Cornwall, 1602*

KERRIER'S coastline stretches from Falmouth in the east round the
Lizard Point and westward to Prussia Cove and the boundary of
Penwith. **Pendennis Castle**, high on a promontory overlooking
Falmouth Bay, guards this large domain. In the 1530s, when Henry
VIII ordered the castle to be built, the headland belonged to the
Killigrew family. John Killigrew was appointed first governor of
Pendennis in 1544 and was knighted that same year. Today, Pendennis
is considerably larger than its counterpart castle of St Mawes across the
water, but they were much the same size originally. Pendennis
consisted of a circular keep which housed the garrison, and a domestic
block at the back where the governor lived. It was surrounded by a dry
moat which was crossed by a drawbridge.

As with all Henry VIII's castles, it fell to the governor to maintain
and equip the garrison at his own expense. The Killigrews were
negligent in their duties in this respect. In early January 1582 a Spanish
ship driven by foul weather into Falmouth haven was obliged to remain
there for several days. The temptation was great. Killigrew's men
boarded her, stripped her of goods and carried her out to sea. Bolts of
holland cloth and leather chairs suddenly appeared in the Killigrews'

Pendennis Castle, Falmouth. Built in the 1530s to strengthen Henry VIII's coastal defences.

home. The full facts of this murky affair never emerged, perhaps because John Killigrew, as governor of Pendennis, led the official inquiry.

Elizabeth I built the bastioned outer defences of Pendennis in the late sixteenth century. These included walls round the headland with embrasures for cannon and a stone-faced ditch in front. Intended to keep out the Spaniards whom Elizabeth so feared, these extra defences proved vital at the end of the Civil War, when Pendennis was besieged for five months by the Parliamentarians. In the end it was starvation which forced Sir John Arundell to surrender, not any weakening of the castle's defences.

THE Killigrews remained in control of Pendennis for well over a century, living at **Arwenack**, a house right on the waterfront overlooking the great natural harbour at the entrance to the river Fal. The first governor of Pendennis owned lands that stretched from the river Fal to the Helford Passage and with the tithes of various parishes brought him a handsome income. In keeping with his status as governor, he pulled down the old house and built a new one in its place.

The new Arwenack was not finished until 1567. John died as the last stones were being laid. Built on three sides round a quadrangle, the fourth side was open to the harbour. A castellated gate tower flanked by low walls protected this vulnerable fourth side from attack by sea. Palisades and earthworks surrounded the three landward sides. At the north corner of the house stood another tower pierced with loopholes for bows and muskets; since the house lay at sea-level, Killigrew was taking no risks. A great banqueting hall filled the central section of the house. Here feasts were held for captains of the many ships that dropped anchor in Falmouth haven, Walter Raleigh among them.*

At the end of the Civil War Arwenack was burned to the ground by its owners in order to prevent the Parliamentarians from using the house during the siege of Pendennis. This selfless act of patriotism did not pass unnoticed, for in 1661 William Killigrew was made a baronet. Arwenack was never wholly rebuilt after the Restoration, although it was partially restored to suit the needs of occasional residence, and today it is possible to see part of the Elizabethan shell. Adjacent to it is a new house which has now been converted into a number of flats. Every effort has been made to recreate the appearance of old Arwenack, including a courtyard with little parterres surrounded by neatly-clipped box hedges. It is easy to imagine the Killigrews watching the ships unload their cargoes on the wharf. After the establishment of Falmouth as a station for the Post Office packet boats at the end of the seventeenth century, all manner of merchandise was seen on the quayside – gold, wine, and fruit from Spain and Portugal, rum and sugar from the West Indies.

When Sir Peter died in 1704 the Killigrew baronetcy became extinct,

Arwenack House, Falmouth. Remains of the Elizabethan manor burned to the ground during the Civil War.

Arwenack House. Part of the recent restoration.

his only son having been killed in a scuffle at a Penryn inn. The Arwenack estates passed to his eldest daughter Frances and a century later became the property of Lord Wodehouse when he married Sir Peter's great-great granddaughter Sophia.

TWO other mansions in Kerrier, Carclew and Roscrow, must be mentioned here. The manor of **Carclew**, which runs down to Devoran creek, belonged to the Daungers family in the twelfth century, but the name died out in the early fifteenth century when the last two daughters married a Renaudin and a Bonython respectively, and became co-heiresses of the estate. The Renaudin line soon petered out, leaving the Bonythons in sole possession of Carclew until 1697 when the last male heir, Richard, died.

Richard Bonython's only daughter Jane married a Penryn merchant called Samuel Kemp, who immediately set about reorganising his wife's considerable assets. He sold off much of her land and started to build a handsome new house with the proceeds. He was never to enjoy the grand home he envisaged for he died in 1728 before the house was completed. Jane, less attached to visions of grandeur, promptly stopped all work upon it. On her death some ten years later she left Carclew to her kinsman James Bonython of Grampound and in 1749 James sold the property to the mining magnate William Lemon.

Lemon was born of humble family in the tiny village of Breage, near Helston, and became the manager of a tin-smelting house at Chyandour, Penzance. He married well, and in 1724 used his wife's dowry to open a tin mine on land belonging to Lord Godolphin at Ludgvan. It became known as Wheal Fortune and was reputed to have earned Lemon £10,000. Thereafter he moved to Truro and began to work the rich Gwennap copper mines. Impressed by the elegant **Princes House** Thomas Edwards designed for him in 1737 in the centre of Truro, he now engaged this talented architect to alter and complete the half-finished house on his newly acquired country estate.

An engraving of Carclew in Borlase's *Natural History of Cornwall* (1758) shows Edwards' design for the house. A two-storeyed rectangular block set on a basement storey to give it extra stature, it is flanked on either side by a low colonnade. Edwards' use of a grand Ionic portico and small pavilions gives Carclew a decidedly Palladian air. The design of the central block resembles that of Wanstead House in Essex (begun in 1713) by Colen Campbell, an engraving of which Edwards could have seen in Campbell's *Vitruvius Britannicus*, published in 1715. Wanstead was praised by devotees of the Palladian movement for its classical order and simplicity, for its clear cut horizontal and vertical divisions and for its magnificent hexastyle portico, 'the first yet practised in this manner in the Kingdom'. The severity of its design was softened by an elegant balustrade adorned with sculpture along the roofline and by contrasting curved and triangular pediments over the

Princes House, Truro. Designed by Thomas Edwards in 1737 for William Lemon.

windows in the manner of Inigo Jones. Carclew had none of these refinements and was unduly harsh in appearance as a result.

William Lemon died in 1760, leaving his descendants money, a baronetcy, and a zest for public life. They lived at Carclew for a hundred years, until the name became extinct in 1864 on the death of Sir Charles Lemon. His nephew Arthur Tremayne inherited the estate and continued to improve it. Unhappily, Carclew was burned to the ground in 1934 and only fragments of the façade now remain. Photographs taken before the fire show that the house rivalled

Carclew, near Mylor. Altered and completed by Thomas Edwards c. 1750 for William Lemon. Engraving from Borlase's Natural History of Cornwall (1758).

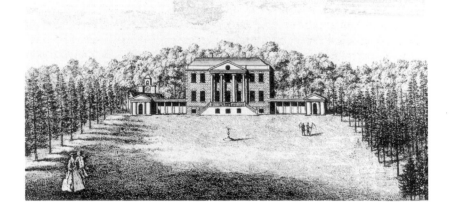

The remains of Carclew, burned to the ground in 1934.

Trewithen and Antony in magnificence. The simplicity of the exterior belied the lavish decoration of the rooms inside. A graceful colonnade of Ionic columns – by now the hallmark of an Edwards interior – made the entrance hall as coolly elegant as a Roman atrium. Both hall and staircase were richly decorated with elaborate plasterwork and a further screen of Corinthian columns on the upstairs landing complemented the colonnade in the hall. The theme was one of classical grandeur, but grandeur tempered by the delicate, flowing curves of a wrought-iron balustrade, making a perfect blend of rococo charm and classical austerity.

Magnificent broad-leaved trees and massive rhododendrons still flourish in **Carclew Gardens**. In addition visitors can see effective new plantings of golden yew, red maple and berberis.

A FEW dilapidated outbuildings are also all that are left of **Roscrow**, a mile inland from Penryn, although a modern house has been built in its place. Those readers familiar with the writings of Mary Delany, the famous eighteenth century diarist who laid bare her soul in her *Autobiography and Correspondence* published a hundred years later, will remember that it was to old Roscrow that she was taken in 1718, as the young bride of Alexander Pendarves, a gouty sixty-year-old to whom she was unhappily married.★

This is how she described the old house:

The castle is guarded with high walls that entirely hide it from your view. When the gate of the court was opened and we walked in, the front of the castle terrified me. It is built of ugly, coarse stone, old and mossy, and propt with two great stone buttresses, and so it had been for threescore years. I was led into an old hall that had scarce any light belonging to it; and on the lefthand of which was a parlour, the floor of which was rotten in places, and part of the ceiling broken down, and the windows were placed so high that my head did not come near the bottom of them.

It would seem that Roscrow had changed little since Tudor times, and still kept its semi-fortified appearance. However, Mary was allowed to fit the house up to her own liking, and for a time the task of modernising and redecorating took her mind off her situation, but she found it hard to be cheerful. The high grey walls of Roscrow made it a prison, separating her from her family and from sanity.

Only when her beloved brother and mother came to stay or when her husband was away could she see Cornwall in a kinder light. Relieved of Alexander's presence, she became aware of the countryside and, despite its associations, admitted that Roscrow was a beautiful spot:

It was placed on the side of a hill [which fell gently from the front of the house] surrounded by pleasant meadows, which by an easy descent opened a view to one of the finest harbours in England, generally filled with shipping. The prospect was enriched with two towns [Penryn and Flushing] one considerably large and a castle [Pendennis] placed on an eminence which at some distance looked like an island. The chief town [Penryn] was a peninsula, and situated on a high hill; it consisted of one large street which crossed the summit of the hill, by which advantage every house had a falling garden and orchard that belonged to it; and what is yet more singular, a rivulet that ran through each. These gardens and orchards entirely covered the hill, so that to every eye which beheld it at a distance the whole appeared a garden, and in great bloom at its proper season. Indeed nothing could be more delightful or beautiful in the month of May or June: the whole terminated in an unlimited view of the sea.

But a view, however beautiful, was not enough to sustain her, and after her husband's death in 1724 Mary never set foot in Cornwall again.

She was left a widow, and not a rich one, for Pendarves had lost part of his fortune in the South Sea Bubble catastrophe and part of it drinking and gambling. Nor had he altered his will, and his nephew Francis Basset inherited Roscrow after all.

BETWEEN Carclew and Roscrow lies another great estate. **Enys**, a Celtic word for an island (it lies at the base of a peninsula between Mylor and Penryn creeks), still belongs to the Enys family who took

their name from the place in the eleventh century. An engraving in
Borlase's *Natural History of Cornwall* shows a large E-shaped
Elizabethan mansion with projecting wings standing on flat ground. A
long, formal garden with two pavilions at the end – one survives as a
garden cottage – leads to woodland on one side, while cattle graze in the
distance. Samuel Enys, Sheriff of Cornwall in 1709, married the
daughter of a rich London merchant. When her two brothers died
without heirs, Dorothy inherited a fortune. She and Samuel enlarged
the house and laid out the gardens. These were famous in the eighteenth
century when Borlase made his engraving.

The present house was built in the 1830s upon the old foundations.
Henry Harrison, who worked at Port Eliot, was the architect, but the
new Enys house was a commission of a different kind. He built a
handsome square block whose simple but satisfying proportions are
the perfect counterpoint to the flamboyant garden that surrounds it.
Parts of the old house – an oak-panelled room, staircase, and windows
– still survive at the rear, but otherwise Enys is Victorian loftiness and
solidity at its best.

Winding paths overhung by magnolias and rhododendrons as well as
rare specimen trees surround the house on three sides, affording
colourful views from all the main reception rooms. The long drive is
flanked for half a mile by a jungle of deep crimson rhododendrons

Enys. Designed by Henry Harrison c. 1836 and built on the old foundations.

Enys, near Mylor. Engraving from Borlase's Natural History of Cornwall (1758) showing the E-shaped Elizabethan mansion.

which rise to over forty feet. On the other side, Pendennis promontory can be glimpsed in the distance. Far below the house, at the bottom of a deep dell that is smothered in bluebells in May, are a series of rectangular ponds fed by a stream. Wild duck live in this gigantic water-garden, and primulas, water-lilies, and yellow iris grow on its banks.

HOUSES and gardens of a different kind, more intimate and informal, are to be found on either side of the Helford river, that slender tongue of tidal water that flows from the mouth of the sea to the little village of Gweek with its two bridges that span the high tide mark. Ancient manors and farmsteads alike were usually built at the steep head of a hidden valley, where level ground might be found. Their gardens sloping down to the edge of a creek are often outstandingly beautiful.

Glendurgan, on the north side of the river across the water from the village of Helford, is one such magic place. The existing house belongs, as it always has done, to the Fox family. Built in the 1840s after the original small thatched cottage had burned down, its owner, Alfred Fox, planted the garden in the 1820s and 30s and designed both the curved drive and walled garden, now filled with roses and fruit trees.

As the chief shipping agents in Falmouth, the Fox family had every opportunity to ask travellers to bring back exotic plants from all parts of the globe. Tender and even sub-tropical plants grow happily in the open in this sheltered climate, and shrubs quickly become trees.

Certain conifers flourish particularly well in a warm, damp climate, and Glendurgan has many of them. Alfred's fifth son, George Henry Fox planted the Deodar and Atlas cedars, the swamp and Mexican cypresses and the weeping spruce, some of whose trunks have twisted into fantastic shapes. The tulip tree, *Liriodendron tulipifera*, competes with the conifer for sheer size, its trunk of thickly knotted bark as fat as the sturdiest oak or beech.

Glendurgan, on the north side of the Helford river. The house was built for Alfred Fox in the 1840s, after the original thatched cottage had burned down.

Glendurgan garden, sloping down to the Helford river.

Half-way down the garden, in a dip between the two steep slopes of the valley, a maze of laurel confuses and delights. Below this is a pond. In early spring arum lilies flower on one side, while on the other the gigantic leaves of *Gunnera manicata* from Brazil spread over the water like huge fans later in the summer. Japanese maples and rhododendrons line the path from the maze to the pond.

George Henry's cousin Robert Fox of nearby Penjerrick raised many hybrid rhododendrons (the original grove of *R. barclayi*, named after his son Robert Barclay Fox, still survives in this lovely but now overgrown garden), which found a home at Glendurgan. The fragrant creamy yellow 'Penjerrick' by the pond is one of the loveliest. On succeeding to the property in 1931, Cuthbert Lloyd Fox planted Asiatic rhododendrons, magnolias, camellias, hydrangeas, and eucryphias in a series of glades down the valley slopes. Brilliant flashes of red and orange catch the eye, glinting like the panes of a stained-glass window. A gate leads to the hamlet of Durgan, a handful of cottages on the quayside and the first point on the Helford river where landing is possible whatever the state of the tide. A steep, leafy lane leads from Durgan along the eastern slope of Glendurgan, and half way up another gate brings you back into the garden.

IN in the mid-nineteenth century Charles Fox laid out another garden called **Trebah**, four miles south of Falmouth near Mawnan Smith. Here a semi-circle of mature trees surrounds lawns which stand above a 200-foot ravine down to the river Helford. As well as handsome examples of rhododendrons, azaleas, pieris and camellias, many sub-tropical species are also to be found. These include dicksonia, cyatheas, woodwardias, echiums, beschornerias and aeoniums. An extensive water-garden has been created alongside the valley stream which cascades over massive granite boulders to form a dramatic series of waterfalls. Moisture-loving zantedeschia, lobelia cardinalis, ligularia, iris, meconopsis and primula thrive near the rock pools. Tree ferns and palms such as *Trachycarpus fortunei* from China add to the sub-tropical atmosphere of Trebah.

Other valley gardens which command fine views of the river Helford are **Carwinion** (near Mawnan Smith) and **Trelean** (near St Martin-in-Meneage). The latter is renowned for its planting of cornus, acers and other shrubs for autumn colour.

TRELOWARREN, the home of the Vyvyan family for over five hundred years, is set in gently rolling parkland to the south of the Helford river. The manor was held by the powerful Cardinans in the thirteenth century but passed by marriage to the Ferrers family in 1279. The

Trelowarren, to the south of the Helford river. Engraving from Borlase's Natural History of Cornwall (1758) showing the east (entrance) front and the two storey-bay (dated 1662).

Ferrers remained at Trelowarren until 1426, when Honor Ferrers married John Vyvyan, whose family came from the parish of St Buryan in Penwith. Indictments for assault, murder, and wrecking show the early Vyvyans to have been a wild and lawless lot, but John's inheritance of Trelowarren brought them respectability and a place in Cornish society. His son Richard married Florence Arundell of Trerice, and in 1544 his grandson Michael was appointed Captain of St Mawes Castle, which was then nearing completion.

Michael Vyvyan died in 1561 and was succeeded by his second son John, who brought vast new estates to the family upon his marriage to a co-heiress of Edward Courtenay, Earl of Devon. He and his bride began to rebuild the mediæval mansion and the walls of the three bays of the chapel that now adjoins the house, including the outer door with its pointed arch and naively carved spandrels and lintel, are all that remain of the earlier building. Some years later, in 1636, Richard Vyvyan secured a licence to restore the chapel and arranged to enlarge the house still further. Work was interrupted by the Civil War in which John Vyvyan's descendant Richard played a prominent part. Charles I commissioned him to coin money (coin being made from plate contributed by Royalist supporters) and he set up mints at Truro and Exeter. He was also in command of a fort erected on Dennis Head to guard the Helford river. Its construction was financed largely by Richard himself, who gave over £9000 of his own money to the Royalist cause. Already knighted after a court masque in 1636, Charles I conferred a baronetcy upon him at Boconnoc in 1644. His wife became Lady-in-Waiting to Queen Henrietta Maria.

Sir Richard completed his proposed enlargement of Trelowarren after the Restoration, which he commemorated by setting up two stone pillars in the quandrangle formed by the chapel and the north wing, the old Tudor portion. These pillars now form the entrance gate. He also built a two-storey bay on the garden side, the date 1662 being clearly

Trelowarren. The west front, with the enclosed Lady's Garden on the left and the chapel on the right.

inscribed beneath the parapet. At some stage an extra bay was added to the east or entrance front, and the large north wing was built in the nineteenth century. Sir Vyell Vyvyan, the 7th Baronet, was responsible for the interior decoration of the old chapel in the Gothic style at the end of the eighteenth century. The interior of Trelowarren is largely Victorian.

Sir Vyell was also responsible for planting the original grounds, but the gardens at Trelowarren are largely the creation of Clara Vyvyan, the second wife of the 10th baronet, Sir Courtenay Vyvyan. Clara had a deep feeling for nature and a passionate love of her home which she conveys in her many books about Cornwall. In *The Old Place* she describes her struggle to keep up the garden after her husband's death in 1941 and following a period of total neglect during the war when Trelowarren was requisitioned. She returned from Bristol to find 'tents pitched, trenches dug, pipe-lines laid all anyhow, nohow' in the midst of 'our wild garden that we call the Pleasure Grounds'. A cess-pit had been sunk in her prize snowdrop patch, and valuable shrubs which had taken her years to establish were either decimated or lost in the undergrowth.

Dismayed but not defeated, Clara and her gardener George, who believed that 'plants and trees is living things and can feel the same as we do', slowly untangled the shrubs and stripped the ivy from the trees. Gradually the Pleasure Grounds - a long, narrow strip of flat ground running from east to west above the garden front - were restored to order, and the paths through the glades cleared. They were tended lovingly until Clara's death in 1976 but are now semi-overgrown once more. The 1662 wing and the chapel are leased to an ecumenical charity called the Trelowarren Fellowship which is responsible for the maintenance of the house, but sadly the upkeep of more than a mile of pathways is beyond its means.

Clara's spirit lives on in the Lady's Garden, a square enclosure to the north-west of the house and bounded on the south 'by a little iron fence separating it from lawn and chapel'. Essentially a summer garden, Clara filled its five flowerbeds, four triangular ones at each corner and one circular one in the middle, with old garden favourites: lilac, yellow aconites, pink tree-peonies, gladioli, roses, and wood anemones grew side by side, while round the walls were scented plants such as sweet verbena, rosemary, and broom. The plants were of little botanical value but to Clara they were close friends.

A FEW miles inland from Mount's Bay, as the long curve of Kerrier's western shore is called, between Marazion and Helston, stands **Godolphin House**, one of the most historic buildings in Cornwall.

Godolphin is hidden among the trees on the lower slopes of Godolphin Hill, where the tin which made the family's fortune was mined. 'No greater Tynne works in all Cornwall than be on Sir William Godolcan's ground', was Leland's comment when he visited Great Work Mine in 1536. Sir William's nephew, Sir Francis Godolphin, who inherited the estate in 1575, brought in two mining experts from Germany to supervise the sinking of shafts and to explain new techniques. The introduction of processes such as wet stamping and blowing meant that the tin was mined more economically and more efficiently. In the wake of such prosperity Sir Francis set about adding to his house.

As it stands today the oldest part of Godolphin dates from about 1475. It was constructed around two main courtyards, the smaller of which still survives. A study of the four sides of the little quandrangle reveals the alterations made over the centuries. The east side is Tudor, to judge by its round-headed windows, and contains the original dining-room. Its superb linen-fold panelling and carved beams and bosses are finely wrought, and between the windows hangs the royal coat of arms of Henry VIII, said to have been presented to Sir William Godolphin for his services at the siege of Boulogne in 1544. The battlemented wall of the south side represents the original main elevation of the house. An archway in this wall has late Gothic panelling on the jambs and once led to a screens passage and an Elizabethan great hall. Behind lay a second larger court comprising bedrooms and service rooms.

Sir Francis added to and altered the west and north sides of the quadrangle. The elaborately carved doorway in the King's Room (so called because Prince Charles is said to have sheltered at Godolphin *en route* for the Scillies in 1644) was made to commemorate the marriage of

Godolphin House, between Marazion and Helston. The east and south sides of the inner courtyard, showing the round-headed windows of the original Tudor dining-room and the Gothic archway that once led to an Elizabethan great hall.

Godolphin House. The west side of the courtyard, with 16th-century mullioned windows. On the right, the colonnade of the north front, built c. 1635.

Godolphin House. Engraving from Borlase's Natural History of Cornwall (1758) showing the colonnaded north front and forecourt.

his son William to Thomasina Sidney in 1604. The colonnade of the north front was the last part to be added and was built in about 1635 by William and Thomasina's son Francis.★

Eight massive Tuscan columns on the outside and six on the inside of the courtyard form an imposing double loggia over the Elizabethan wall and entrance gateway with its heavy oak doors. These granite pillars carried a whole new range of entertaining rooms with fine views over the forecourt. An engraving by Borlase made in 1758 shows two corner pavilions and rows of trees planted in front of them.

The Godolphin family only lived intermittently at their Cornish seat and in 1850 a large part of the house was demolished. The remainder was leased as a farmhouse until it was sold in 1929. The present owners bought the property over fifty years ago and have spent much time and energy preserving it.

THE land surrounding **Pengersick Castle**, on the outskirts of modern-day Praa Sands, bordered that of the Godolphins. In the fourteenth century Henry de Pengersick, known as 'Le Fort', married Engrina Godolphin. Records testify to his strength and fiery disposition; he was excommunicated from the church for attacking the vicar of Breage and a monk from Hayles Abbey, when they came to collect their dues. On his death in 1343 Pengersick passed to his daughter Alice, and again, through the female line, to the Worth family and then the Milliton family from Devon.

Towards the end of the fifteenth century Pengersick was rebuilt as a fortified Tudor manor, the remains of which still stand today. The letter W worked into the moulding round the uppermost window in the tower suggests that the construction took place in Thomas Worth's time. A drawing made by William Borlase shows the original design of this building viewed from the main courtyard. Because it was situated so close to the sea and in a low-lying position, Pengersick was

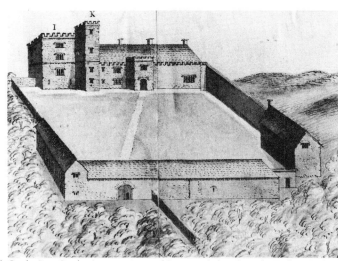

Pengersick Castle, Praa Sands. A reconstruction of the original plan, drawn by William Borlase.

vulnerable to attack from marauding buccaneers. The design of the tower, with its four square rooms one above the other reflects the need to have some defence. A 'drop-slot' was inserted over the entrance doorway so that boiling oil or molten lead could be poured on the unwelcome intruder, and access to the entire house was through one small first floor door. This doorway now connects the tower with the modern wing adjacent to it. A broad newel stair leads up to roof level, and a small battlemented turret - a convenient place to store weapons in time of crises - covers the door onto the leads. From the roof the occupants could survey the countryside from all sides and warn the household of the enemy's approach. The gun room, just below the ground floor level of the tower, is fitted with six arched embrasures which have unusual dumb-bell shaped gun-loops. The enemy had little chance of ascending the tower and gaining access to the rest of the house. ★

The first John Milliton's grandson William married Sir William Godolphin's daughter Honor. She arranged for the panelling of the main room to include a series of verses carved in the wood. Both panelling and verses have long since disappeared (although fragments were found in nearby outbuildings), but Borlase made a copy of the verses which has been preserved.

> When marriage was maid for vertew and love
> There was no devorse goddis knote to remove
> But now is muche people fallen into such luste
> That they do break Goddis wyll most juste
> Wherefore unto alsuch let thys be sufficient
> To keipe God is lawe for feare of his punishment
> In the burning lake where is most of all torment.

Tragedy befell Pengersick when William and Honor lost their only

Pengersick Castle. The late 15th-century fortified tower.

son at sea. When William died in 1571 the property was left to their six surviving daughters. As often happens when an estate is divided into impractical units or when daughters have husbands with properties of their own the inherited property is neglected. Pengersick Castle quickly fell into disuse. In time the fortified tower decayed and the materials were used to build cottages nearby. Mouldings from one of the castle doorways can be seen round the front door of the adjacent farmhouse.

By the time the Duke of Leeds's properties came on the market in 1922, Pengersick had become part of the Godolphin estate. The castle was little more than a hollow shell throttled with ivy and used as a shelter for cattle. The purchaser restored the tower and built a small modern wing beside it. An effort was made to recapture the atmosphere of a Tudor garden. Trees were planted destined for topiary work, and a series of rising terraces led to a reconstruction of the original arched entrance to the outer court as seen in Borlase's drawing. The present owners completed the restoration work and cherish both Tudor tower and garden.

No picture of Kerrier would be complete without reference to a smaller, more modest type of manor house, such as **Truthall** which stands high on a hill plateau behind Helston. In the Middle Ages it belonged to the Nance family, and the original low, rectangular hall-house, now part of the farm buildings, can still be seen. A cross passage runs from front to back with the hall, open to the roof, on one side and a smaller service room on the other with an upper chamber or solar

Truthall. The 17th-century wing built at right angles to the rear wall of the mediæval hall-house (on the right).

above. The arch-braced collar beam roof and the decorative windows of the hall, with their cusped heads and saddle bars, are remarkably ornate for such a simple house and were surely designed to impress. The entrance doorway, too, is imposing with its heavy roll-moulding round the arch and framing hoodmould. The coarse rubblestone wall and rear doorway of roughly hewn moorstone at the back of the house, in contrast, show no such refinement and make no concession to beauty.

In the middle of the seventeenth century a more up-to-date house was built at right angles to the mediæval dwelling, which was demoted

Truthall, near Helston. A window of the mediæval hall-house, now part of the farm buildings.

to provide further domestic quarters. The date 1642 is inscribed over the front door, which now has a flat head instead of an arch. The ground floor consisted of two rooms, the front hall and another room behind, while upstairs was a fine parlour with mullioned windows and a fireplace. A handsome nineteenth-century house with light, spacious rooms and wonderful views was built beside the seventeenth-century addition.

9.

Penwith

Majestic Michael rises; he, whose brow
Is crowned with castles and whose rocky sides
Are clad with dusky ivy; he whose base,
Beat by the storm of ages, stands unmoved
Amidst the wreck of things - the change of time.

Sir Humphry Davy

ST MICHAEL'S MOUNT, standing proudly on a rocky island in Mount's Bay, changes its mood like magic according to the time of day, the time of year. On a stormy winter's day when the wind howls and whips up the waves to a white foam, the Mount is a forbidding fortress which has withstood three sieges and stands ready, still, to defend Penwith. Yet on a warm summer's evening the island is bathed in a pink glow, a vision of welcome to pilgrims who have travelled from far and wide to worship at the shrine of St Michael.

In the late Bronze Age Penwith was the most highly populated part of Cornwall and St Michael's Mount an important port and trading centre. Ireland, rich in gold and copper, needed Cornish tin to make bronze axe heads, cooking and storage vessels. Rather than risk a voyage round Land's End, feared for its rough seas and treacherous currents, Irish traders landed their cargoes in the Hayle estuary on the north coast and proceeded overland across gentle hills to St Michael's Mount on the south coast. Natives provided horses to transport the wares and acted as guides over the rough tracks, following the route of the present-day A30 road from Hayle to Penzance.★

St Michael's Mount c. 1880, before the trees were planted.

Domesday Book contains some evidence that there was a religious community on the Mount in Saxon times, but the earliest record of a monastic settlement dates from 1135 when Bernard Le Bec, Abbot of Mont St Michel in Normandy, built a Benedictine priory high up on the rocky island crag. The foundation was made possible by the grant of lands in Cornwall by Robert, Count of Mortain, William the Conqueror's half-brother, to Mont St Michel. Throughout the Middle Ages pilgrims came to the shrine of St Michael to make a vow or as part of a penance required of them by their confessors. ★★

In 1275 an earthquake destroyed the original priory church. It was rebuilt in the late fourteenth century and is still used for public worship. Although extensively restored in the nineteenth century, the simplicity of the interior faithfully reflects the austere existence of the monks. The stone tracery of the two beautiful rose windows is original, but the stained glass is Victorian.

When Henry V went to war in France the priory was seized by the Crown as an alien religious house, and in 1424 it became part of an endowment for the Brigittine Abbey of Syon at Twickenham. Thus ended any connection with Mont St Michel.★

After the Dissolution of the Monasteries in 1536, the Mount was leased by the Crown to successive members of the local gentry who acted as governors and were required to maintain a garrison. Its subsequent history was chequered, until its Royalist owners sold it in 1659 to Colonel John St Aubyn, whose descendants live there today.†

St Michael's Mount.
The Chevy Chase room.

St Michael's Mount. The Blue Drawing-room.

In 1954 the 3rd Lord St Levan handed the Mount over to the National Trust. It is now the home of his son, the 4th Lord St Levan.

Today tourists flock across the causeway at low tide or pile into small boats at high tide, to visit a romantic, fairy-tale home. Despite the addition of a Victorian wing below the original buildings, the external appearance of the castle and chapel has hardly changed since the St Aubyns took possession. The library is in the oldest part of the castle, its two small sash windows ingeniously fitted into the lower portions of the twelfth-century lancet windows. It is easy to imagine the monks staring through the narrow slits out to sea, offering silent prayers perhaps for a brief respite from the merciless wind. The Chevy Chase room is named after the seventeenth-century plaster frieze depicting hunting scenes based on the mediæval *Ballad of Chevy Chase*. The hounds sniff furiously and rabbits cock one ear or emerge timidly from their burrows. The room was originally the monk's refectory, the chairs being nineteenth-century copies of those used by the monks. Shields bearing the coats of arms of prominent Cornish families hang on the walls, for the St Aubyns were linked by marriage to the Bassets, Grenvilles, and Arundells. The first St Aubyn owner of the Mount, Colonel John St Aubyn, married the second daughter of Sir Francis Godolphin of Treveneague, and over the castle entrance are the arms of the St Aubyn family impaling Godolphin.

The ruined Lady Chapel was converted into the present Blue Drawing-room between 1740 and 1750 by the 3rd baronet, a Member

St Michael's Mount. View from the south-east showing the terraced garden.

of Parliament with a reputation for incorruptibility. Incorruptible may be, but compulsively in tune with the fashions of the time, he created an elegant interior in the latest fanciful 'Gothick' style with 'Gothick' Chippendale chairs to match - all in pastel blue and white. Cornish gentry may have been aware of Horace Walpole's Strawberry Hill, a fashionable dilettante haunt being decorated at the same time, but few would have had the chance to visit it. No doubt local families longed to be invited to some entertainment at the Mount in order to inspect and pronounce an opinion on the new style - so light and frivolous compared to their own dark-panelled drawing-rooms with sober, sensible Queen Anne furniture.

The 3rd baronet was also a practical man determined to help the local inhabitants. In 1727 he rebuilt the Mount harbour in order to stimulate the export of Cornish tin and copper. A hundred years later fifty-three houses had been built and a thriving fishing industry established. Most of these cottages were pulled down in the early twentieth-century and replaced by the two rows of cottages standing today. A small herd of cows was kept on the island in Victorian times, the milk made into butter and clotted cream in the octagonal dairy, which the visitor passes on the way up to the castle. Stores and fuel are still taken up by means of an underground railway built at the turn of the century.

On the north, more protected and landward side of the island a natural rock garden surrounds the steep path up to the castle. Sturdy pines, evergreen oaks and sycamores manage miraculously to withstand the wind and give shelter to mature camellias, azaleas and rhododendrons. These shrubs owe their existence to the trees, for without a buffer they would be flatened or uprooted by the relentless wind. New shelter belts are desperately needed to secure the future of the garden in years to come, but considerable thought goes into the buying and positioning of new trees. The Japanese *Pinus Thunbergii* has shown itself to be the most wind and salt resistant. The present Lord St Levan hopes to establish a new type of Dutch elm apparently immune to the dread disease.

Some of the terraces on the eastern side of the island were laid out two hundred years ago. Today these tiny terraces which face south contain an astonishing variety of plants. Different types of aloe adorn the precipice below the castle, the lovely orange *aloe arborea* flowering in February. In the summer mesembryanthemums, ivy-leaf geraniums, and blue convolvulus dangle over the rocks and fill every crevice.

The series of miniature walled gardens formed by each level of terraces are full of surprises. As you look down at them from the castle battlements, the sea, not the rock face, is the backdrop to the gardens, its colour changing from midnight blue to peacock according to the colours of the sky. Muted Cezanne greens come next to blues, for the grass grows to the water's edge, a broad swathe being cut to form a path right round the island. The monks are thought to have established the sweet-smelling Mount lilies (a type of narcissus) which cover the grassy lower slopes in spring, to be followed in summer by masses of blue agapanthus and red hot pokers.

Above this semi-wild domain the terraces become more formal, more man-made, as if to deny or defy the power of the sea. Veronicas, shrub roses, marguerites, pelargoniums, and wall flowers struggle to keep their petals behind hedges of escallonia and fuchsia, while higher up, in a middle zone which the wind somehow by-passes as it wages war on the castle above, a number of rare and tender plants bloom relatively unperturbed. Among the most remarkable are the blue passion flower *passiflora caerulea, bigonia grandiflora* and *bigonia cupensis, stauntonia latifolia,* lion's paw from South Africa *leonotis leonurus, sparmannia africana* and *clematis flammula.* Now the planting continues. Lord St Levan, his head gardener Mr Bowden and their adviser from Tresco, Helen Dorrien Smith – a triumvirate as tough and tenacious as the plants – direct proceedings. Their optimism, determination and imagination have protected and embellished the most unusual garden in Cornwall.

THE truly spectacular gardens at **Trengwainton**, five miles away on the mainland, are an extraordinary contrast to the windswept Mount and show what a gardener's paradise Cornwall can be. Originally belonging to the Arundells of Trerice, the property was bought in 1814 by Sir Rose Price, the son of a wealthy Jamaican sugar planter. Price planted the trees, mainly beech, ash and sycamore, which now stand round the house and grow beside the drive to the lodge and entrance to the gardens. This mature wood is a beautiful natural setting for all the shrubs subsequently planted, and provides the shelter so conspicuously absent on St Michael's Mount. It is an exotic and enchanted world.

William Wilberforce's Emancipation Act of 1833, which abolished slavery, ruined the Price family's fortunes, and the estate was sold in 1867 to T. S. Bolitho, a banker with substantial mining interests. His son, Thomas Robins Bolitho, enlarged the hitherto modest house to its

Trengwainton garden. The drive.

present grand proportions in 1897, and built the wide carriage drive. His nephew, Edward Bolitho, inherited Trengwainton in 1925, and had the daunting but exciting task of creating a garden. He sensibly sought the advice of other Cornish gardeners, notably J. C. Williams of Caerhays who gave him rhododendrons and other shrubs. George Johnstone of Trewithen then suggested that Bolitho should take a share in Kingdon Ward's 1927-8 plant-collecting expedition to north-east Assam and the Mishmi Hills in Upper Burma. Many of the huge rhododendrons and magnolias were raised from seed gathered on this expedition.

On the right of the drive are a series of inter-connecting walled gardens remarkable not only for the amazing variety of the shrubs, but also for the sheer size to which they have grown. In February and March the boughs of tall, graceful magnolia trees from Japan, Sikkim, China, and Tibet are laden with pure white, pale pink, rose-coloured, or purple blooms. One by one the huge but fragile petals flutter onto the path or grass below. Sweet-smelling eucalyptus and acacia from Tasmania, *Styrax japonica* from Japan, with its white, bell-shaped flowers, evergreens from South America and the Himalayas, passion flowers, fuchsias, azaleas, camellias, and rhododendrons are carefully nurtured within these walls.

Further up the drive is a small meadow planted with trees in honour of the royal family - a fine oak commemorates Queen Victoria's Diamond Jubilee in 1897, a lime tree Edward VII's coronation in 1901. Queen Elizabeth the Queen Mother planted a Bhutan pine with long bluish-green needles in 1962 and Princess Anne a Mexican pine in 1972, both large trees now.

Magnificent rhododendrons - white, scarlet, deep crimson, magenta pink and yellow - flank the drive and lawns round the house, rivalling beech and sycamore in height. Beyond the lawn in front of the house is a view of St Michael's Mount framed by dark pink and vermilion rhododendrons. This extensive, open vista gives Trengwainton an airy, informal elegance. Situated in a part of Cornwall known for its moors and stony hills, where only furze and bracken grow, this ninety-acre garden is a luscious oasis of rich reds and pinks and shiny, succulent greens.

ROSCADGHILL near Newlyn, the largest fishing town in Cornwall in the nineteenth century and famed for its school of artists, is a fine, late seventeenth-century house with a neat, symmetrical façade. A Cornish (granite) version of a redbrick William and Mary house complete with small, circular lawn and gravel sweep in front, the whole is enclosed by its own wall. The pedimented front door and well-proportioned sash windows on either side show a sound understanding of the new classical concepts of architecture based on symmetry and proportion. The wealthy gentry were the first to respond to the new ideas, but by the late seventeenth century the movement had begun to influence the design of modest houses, and the circulation of pattern books enabled country builders to work in the new style. The sash windows at Roscadghill are not yet flat-headed but are delicately rounded at the top, a reminder of curves and arches soon to disappear altogether.

Nearby, on gently sloping ground above Newlyn, stands **Trereife**, a house of handsome proportions rebuilt by John Nicholls, a successful Middle Temple barrister, in the early eighteenth century. Having seen the fine Queen Anne houses being built in London with regular brick or stone faces, Nicholls understandably wanted the main elevation of his own home to be as elegant as granite would permit. He thus took the trouble to face the east front with squared rough-cut brown granite and he used the same brown granite for the arches over the window openings elsewhere. The sliding sash window was at the height of fashion and therefore used for the east front whereas casement windows with heavy wooden mullions were used for the west and north sides of the block. Nicholls also raised the first floor of the earlier house to make the two storeys of equal height and put in a new roof with dormer windows. The result is a classical composition of seven bays, the central bay enhanced by a pedimented porch with Doric columns.

The desire for symmetry is also reflected in the original lay-out of the garden. Contemporary engravings in the dining-room show ladies and

Roscadghill, Penzance. A Cornish granite version of a William and Mary house.

gentlemen strolling along straight paths which formed two neat rectangles in front of the house. These are flanked by avenues of trees. The clipped yew hedge that virtually hides the south front was planted in 1780.

Today the garden is softer in appearance. A flat lawn in front of the house leads the eye to the fields beyond where horses graze contentedly. On one side, on a raised terrace, a wide herbaceous border runs the length of the original walled kitchen garden, now used as a paddock. At the bottom of the bank is an equally long border devoted to old-fashioned roses. On the other side of the lawn is a woodland shrubbery filled with mature camellias and rhodendrons. A bulb farm provides many varieties of daffodil for the garden.

The Nicholls family continued to live at Trereife until the end of the eighteenth century. In 1799 the widow of William Nicholls married the Revd Charles Valentine Le Grice, the eldest son of a Norfolk clergyman. 'C.V.', as he was affectionately known by the family, was educated at Christ's Hospital, London and Trinity College, Cambridge. Urbane, witty and intellectual, he thrived in the company of his school friends Samuel Taylor Coleridge, Charles Lamb and Leigh Hunt. Those who knew him assumed he would pursue a literary career. Shortly after taking his degree, however, Charles Lamb noted that 'Le Grice went to Cornwall, cutting Miss Hunt [Leigh Hunt's

Trereife c. 1900.

sister] completely'. Was it to remove himself from an awkward situation or merely the pressing need to earn his living that caused 'C.V.' to accept a post as tutor to the son of the widowed Mrs Nicholls of Trereife?

Whatever the circumstances surrounding his arrival, C.V. proved to be the ideal successor to the Nicholls family. He ran the estate efficiently and played an active part in local affairs, both as a magistrate and as a curate. He died in 1858, having outlived his wife and stepson by more than thirty years. His only son, Day Perry, inherited the estate. His great-great grandson Timothy Le Grice lives at Trereife today.

FARMERS, like the gentry, were not immune to the pressures of fashion. **Trenow**, high on the slopes above the picturesque village of Gulval, is a simple farmhouse of thatch and cob rebuilt in the eighteenth century with all the latest architectural features – at least on the outside. The façade is symmetrical, the roof hipped, and the windows have the new sash frames. But the central front door opens, not as at Roscadghill, into an entrance hall with elegant reception rooms on each side, but into a large kitchen. A small parlour leads off to the right, while the whole of the other side was in fact a cow house with a barn above, reached by stone steps at the back. The façade was not allowed to interfere with practical needs and disguised the real nature of the building.

Trenow, Gulval. A Georgian farmhouse of thatch and cob.

Gulval, c. 1870. A thatched cottage.

CORNISH farmsteads and cottages present a mixture of styles and are difficult to date. The sites of these sometimes lonely dwellings, whether moorland hollow or sheltered valley, were probably selected at the beginning of the Christian era when early Celts began to leave their hilltop strongholds to settle further down the slope and to

Gulval, c. 1870. Rubblestone cottages.

cultivate the land. These early farmers reared cattle, sheep, goats, and pigs, heaving the granite boulders from the land and building stone walls around the tiny fields they tilled.

The Celts continued to live in their humble homes even after the Saxons had replaced the Celtic form of clan ownership with their own feudal system. Today, driving over the bleak hills towards Zennor and the sea, you pass scattered, dilapidated-looking farm settlements – Bosulval, Carnaquidden, Kerrowe – dating from Celtic times. The windy, narrow roads, once rough tracks linking homestead to homestead, sometimes bisect the farm whose buildings are huddled togather on each side. Some of the farmhouses were rebuilt or altered in the seventeenth or eighteenth century when the owners prospered, but the barns and outbuildings have a mediæval air. Cob – a mixture of clay, chopped straw, and even slate rubble which bound together the rough, smallish stones that a man could lift himself – was a cheap and far less durable alternative to granite and few early examples of peasant farmhouses have therefore survived, although the walls of a mediæval house will often have been incorporated into a later farmhouse. The thirteenth or fourteenth century long-houses of turf, stone, and wattle remain part of Lanyon farm, near Morvah, where a low partition separates the byre from the main room. The front entrance was shared by man, woman, sheep, and cow.

The engine-houses of deserted tin mines are the only companions of these lonely farmsteads. Standing desolately on the rugged cliff's edge or high on the moor, they punctuate the barren landscape from St Just to St Agnes, their tall chimney stacks, like Tuscan campanile, silhouetted against sky and sea. Tin-streaming was flourishing again in the first half of the fourteenth century, bronze church bells and cannons, among other things, being in high demand. New trade routes had opened up, not only through the Cinque Ports to Marseilles, but via Bruges to Venice, Constantinople, and the Near East. Cornish seamen shipped the tin in small boats which hugged the coast to London, where it was transferred into French or Italian ships for the channel crossing. In 1305 mining areas were divided into tracts of unenclosed land known as stannaries and a coinage town allotted to each. The area round St Agnes was part of the Tywarnhayle stannary, while Kerrier and far-flung Penwith combined to form another stannary, with Helston as its coinage town. Twice a year, files of packhorses laden with blocks of tin would plod slowly over the criss-cross moorland tracks towards Helston, where merchants, tradesmen, and pewterers assembled to watch the royal officers weighing and stamping the tin.

In early times, the lord of the manor merely received a toll from any tinner or blower who wanted to work on his land. Profit or loss was the tinner's affair. Few of these mediæval lords, although they boasted coats of arms, had houses of any real size or grandeur. The remains of a

Tehidy, near Redruth. Engraving from Borlase's Natural History of Cornwall (1758) showing the elegant mansiond esigned by Thomas Edwards c. 1750 for the Basset family.

house belonging to the Godrevy family in Gwithian, north of Hayle, shows how humble the homes of the minor gentry were. Excavations show a dwelling hardly more than a single long room about thirty feet long by twelve feet wide divided by a wooden screen, one side allocated to the family, the other to the livestock. In about 1400 this hovel was modernised and given a stone wall instead of a screen. The sheep and cows were banished to a separate barn and their shed converted into another room with a solar above, reached by a ladder from the hall. This simple abode was the manor house of **Crane Godrevy**. Early in the sixteenth century sand began to pile up and blow into huge mounds and the house was abandoned.

THE eighteenth century was a different story. Some landowners became fabulously rich when tin or copper was discovered on their estates. Two of the largest mines, Cook's Kitchen and Dolcoath, were controlled by the Basset family. A profit of £7,040 was said to have been made in one exceptional month, providing ample funds to build a grand country house on the site of their hitherto modest seat at **Tehidy**, near Redruth. Thomas Edwards, the architect so popular with the Truro mining magnates, was engaged in 1734 to design the new building.

An eighteenth-century engraving of Tehidy shows a huge, square Georgian mansion with a pedimented central block and four detached angle pavilions, each with a cupola and clock. No expense was spared. Quantities of brick as well as cargoes of Portland stone, yellow Bath stone and even Purbeck squares were brought by sea to Hayle and transported to Tehidy in two-wheeled wagons. A park was landscaped and the stream flowing along the valley to the sea at Godrevy was fed into an excavated area to the south of the mansion to form a lake. Cascades were built in the river outlet and boats were provided for the family and guests. Grandiose in conception, Tehidy was a suitably

Tehidy. 19th-century engraving of the garden front.

elegant residence for a family whose lineage went back to Reginald de Dunstanville, Earl of Cornwall in the twelfth century.

John Pendarves Basset did not live to see his new house completed. He died of smallpox in 1739, aged twenty-five, leaving his widow Ann with £100,000 from mineral profits and rents. His brother Francis inherited Tehidy and his son Francis was created Lord de Dunstanville in 1796. A handsome portrait of him by Gainsborough hangs in the blue drawing-room at St Michael's Mount. Dressed in a ruffled shirt and mustard yellow waistcoat, he looks a cultured, sensitive man who one can imagine entertaining in style.

Lord de Dunstanville died in 1835 with no male heir, and the estates passed to a nephew. The mansion remained unchanged until 1863 when John Francis Basset, whose dues from his various mines were yielding £20,000 a year, decided to enlarge it. Italian artists were commissioned to paint and gild the ceilings of the new drawing-room, dining-room and billiards room, while new servants' quarters were added in the basement beneath. For a few short years life was carefree and glamorous. Then fate dealt a series of blows from which the Bassets never recovered. Mental illness had dogged the family ever since the marriage of Francis Basset to Margaret St Aubyn in 1756. John Francis Basset died in 1869, his brother Arthur committed suicide as his father had done in 1870 and his brother Walter was declared insane. One remaining brother, Gustavus, produced an heir, Arthur who celebrated his coming of age in 1894 with a magnificent ball held in a large pavilion erected for the occasion. Its walls were covered with cream-coloured fabric over which were draped curtains and swags of terra cotta, electric blue, crimson and gold materials hung between large mirrors. This dazzling event was to be the last of its kind at Tehidy.

High taxation, a reduced income from a depressed mining industry and Arthur's unbridled passion for horse-racing and betting made it increasingly difficult to maintain Tehidy and the shortage of ground staff during the Great War precipitated the downfall of this once great estate. Arthur moved to Henley Manor in Crewkerne and the following year, in November 1916, the estate was divided into separate lots and sold. The house became a hospital for the treatment of tuberculosis in 1919. A month after the first five patients had been admitted a terrible fire virtually destroyed the mansion. Only the square angle pavilions of Thomas Edwards's house survive, together with a number of outbuildings. The modern porticoed building dates from 1919. Only the long, wide approach to the house conveys the property's former splendour.

MORE modest houses are to be found in the lush green valleys between Penzance and Land's End. For centuries farmers worked the small arable and grass fields which go right down to the cliff's edge and are protected by stone hedges now covered with great clumps of yellow gorse and a myriad of wild flowers. With the advent of the railway in the mid-nineteenth century, many of these fields were turned into market gardens which supplied not only vegetables but flowers for the cities 'up-country'. In springtime these fields were ablaze with every kind of daffodil, blue and red anemones, and violets, each valley transformed into a huge and glorious garden. Today, because transport and labour costs have soared, it is no longer economical to pick and sell

Trewoofe, Lamorna c. 1920. The central doorway surmounted by the arms of the Trewoofe and Lavelis families.

flowers. The daffodils have to compete with the steadily encroaching bracken and brambles which threaten to smother them completely. For the moment, however, delicate white, sweet-smelling narcissus and golden double daffodils are left to bloom in semi-wild state, and when viewed from afar become a stunning patchwork quilt of every shape and shade of white, yellow and orange.

Trewoofe farm in the valley of Lamorna was once surrounded by such fields and orchards. In the sixteenth century Joanne de Trewoofe and her husband Thomas Lavelis enlarged the house into an E-shaped Elizabethan manor, with a richly sculptured central doorway surmounted by the arms of both families to commemorate their marriage. Their heirs lived at Trewoofe until the end of the seventeenth century when the last male heir was drowned while up at Oxford. By 1706 the manor and its lands had been split into three parts and sold to local families. The west wing was pulled down and rebuilt in the old herb garden; the east wing, though altered in the eighteenth century, still stands. The central portion, with its magnificent doorway, suffered because insufficient land went with it to make it economically viable. It was sold in a derelict state early this century, but its present owner's love and care ensure its preservation.

PART of a fifteenth-century manor house belonging to the Keigwin family still stands in Mousehole. Its deep porch of rather crudely fashioned granite columns surmounted by an upper storey dominates the tiny cobbled street. In 1595 four Spanish ships suddenly appeared off the coast and landed 200 men who razed Mousehole to the ground. Only **Keigwin Manor** survived the onslaught, being built of stone and not cob. Jenkin Keigwin fought to the death to defend it.

THE village of Mousehole, a tangle of narrow, twisting streets which climb steeply upwards from the sea, boasts a proper harbour wall and small wharf. **Penberth Cove**, a smaller fishing community, is tiny in comparison and completely different in atmosphere. Most of the cottages in the wooded valley and on the cliff go back to the seventeenth century and one is still thatched. No modern houses spoil the old-world feeling, for the National Trust ensures that no new building takes place.

The cove itself is a natural harbour. Steeply rising cliffs on either side shelter the inlet, and the boats are hauled safely to the top of a tiny slipway where the fish is weighed and packed.

From the eighteenth and nineteenth centuries huge shoals of pilchards arrived off the shores of Penwith in August and September. Huers, whose task it was to sight the shoals, shouted directions in front of their cliff-top hut to the men in their boats below. The pilchards were netted in long, cumbersome seine nets and dragged ashore to be

Mousehole c. 1880. Children playing outside a 15th-century manor house that belonged to the Keigwin family.
Penberth Cove c. 1880. The pilchard cellars.

salted and packed. The men heaved the fish out of the boats with wooden shovels and carted them in barrows to the pilchard cellars, which can still be seen to the left and right of the slipway. They resemble a row of cottages from the outside but, in fact, the ground

Penberth Cove today.

floor of each is a cellar. When the doors are open, you can see the floors of black elvan stone on to which the pilchards were tipped. Elvan is even harder than granite and much finer in texture. Large slabs were brought by boat from the Lizard and slotted in rows, end-up like tombstones, to make a deep, non-porous, long-lasting floor. Stacks of pilchards, each layer carefully salted, remained in the cellar for five or six weeks oozing oil (used for cooking and lighting lamps), which trickled along wooden gulleys in the floor and down into wooden barrels.

Women came from as far afield as St Just to press and pack the pilchards for export in barrels to Italy and other Mediterranean countries. All pilchard profits went to the squire or parsonwho had set up the company and supplied the boat and nets. The huers and seinesmen who did the work and braved the weather were merely paid for their services. Few were fishermen; they were mostly farmhands or miners taking time off.

In the early 1900s the pilchard shoals disappeared and never returned. Penberth men turned to lobster and mackerel fishing and grew daffodils, anemones and violets in the valley in order to make ends meet. The sons of the fishermen learned a trade to safeguard against hard times. Today, when unemployment in trade is high, they return to fishing. Whereas twenty years ago there were only three full-time boats, now there are eleven. The fish is sold by individual boat but the men have devised a cooperative system of fishing. They need each other to push the boats into the water and so the boats always go out and return together. They rely on the 'Captain' to work the electric winch which pulls the boats back up the slipway and they take it in turns to drive the lorry, bought collectively like the winch, into Newlyn where the catch is sold.

EARLY this century the lane leading down to the cove consisted of rough, granite stones worn into position by carts trundling up and down the valley. This tiny road still follows the swiftly flowing brown-bedded stream that tumbles into the sea by the slipway. Half-way down to the cove, the stream is bridged by a solid-looking gatehouse. This is the entrance to **Foxstones**, an early twentieth century house built in a long, low semi-circle facing the sea.

The drive sweeps up to the house in a graceful arabesque, past a natural rock garden crammed with dwarf azaleas, rhododendrons and hydrangeas, the stream forming a natural boundary between the garden and the road. Designed to enjoy the presence of the sea yet to suggest a retreat from it, it is from the verandah that the setting of this exquisite garden is best appreciated. The lawn below slopes gradually down to the cove, narrowing finally to a grassy path beside which are the ruins of an old mill-house. Called Melynon, meaning the mill in the valley, it is mentioned in King Athelstan's charter, dated 943, to the monks of St Buryan. Shrubs of every kind flank the little paths round the lawn but camellias take pride of place. There are some forty species at Foxstones, the dark crimson *C. japonica altheaiflora* being one of the most beautiful.

ANYONE who has read Daphne du Maurier's *Rebecca* will remember Manderley, the great house to which Maxim brought his shy young second wife. They motored down from London in early May, 'arriving', as Maxim said, 'with the first swallows and bluebells' and turned into the drive which threaded like a ribbon before them.

Above our heads was a great colonnade of trees, whose branches nodded and intermingled with one another, making an archway for us, like the roof of a church. Even the midday sun would not penetrate the interlacing of those green leaves, they were too thickly entwined, one with another, and only little flickering patches of warm light would come in intermittent waves to dapple the drive with gold.

I sometimes wonder if Daphne du Maurier might have had **Boskenna** in mind when she wrote *Rebecca*, for the approach to it fits the description so perfectly. Two sturdy granite pillars guard the entrance to the drive, which runs in a straight line deep into the heart of a wood. In winter-time, trees shudder and creak in the wind, their trunks and stunted branches bent pathetically towards the land, as if struggling to escape from the relentless gales. Hundreds of daffodils and primroses line the drive in March, and in May the trees celebrate

Boskenna. The drive.

Boskenna. The garden front. The gable on the left is mid-17th century (restored in 1858) and the gable on the right was added in 1886.

their survival by forming a canopy of green over a carpet of bluebells. Huge turquoise blue and white hydrangeas flourish in the autumn shade.

The drive bends in the distance and turns again. Round the last corner, trees give way to shrubs and the approach broadens into a gravel sweep with a simple granite sundial in the middle. Solid and rambling, Boskenna is not quite Manderley; its modesty would be out of place in a romantic novel. The square porchway, designed to keep out wind and wet, is unobtrusive, its smoke-grey granite walls starkly bare. The irregular roof-line, proof of later additions to the seventeenth century core, gives the house an informal air. It belonged to William Carthew in the sixteenth century and passed to the Paynter family in the seventeenth. Paynters lived here for 200 years until the 1950s when to pay off huge loans, the estate was broken up and the farms sold.

Lawns flanked by large borders, a colourful hotch-potch of flowering shrubs and perennials, lead round to the other side of the house, more uniform in shape with its two gable ends and tall mullioned windows. Outside the library and drawing-room a path runs between two rows of yellow and pink climbing roses and through the trees to the glistening sea beyond. Garrulous rooks cackle noisily, but do not disturb the peace of Boskenna, which was my childhood home and where my love of Cornwall, its people and traditions, its wildness and poetry, its houses and gardens began.

Bibliography

I am grateful for the use of material from articles by Christopher Hussey, John Cornforth, Mark Girouard and others in *Country Life* and by Michael Trinick in the Royal Institute of Cornwall Journal and a number of National Trust publications.

I have consulted the following books:

Borlase, W. *The Antiquities of Cornwall* 1754
  The Natural History of Cornwall 1758
Brendon, Piers *Hawker of Morwenstow* Jonathan Cape 1975
Carew, Richard *The Survey of Cornwall* 1602
Chesher, V. M. & F. J. *The Cornishman's House : An Introduction to Traditional Domestic Architecture in Cornwall* D. Bradford Barton, Truro 1968
Clarendon, Edward Hyde, Earl of. *The History of the Rebellion and Civil Wars in England* (Ed. W. D.Macray) 1888
Coate, Mary *Cornwall in the Great Civil War* D. Bradford Barton, Truro 1933
Delany, Mary *Autobiography and Correspondence of Mary Granville, Mrs Delany* (ed. Rt. Hon. Lady Llanover) 1861
Du Maurier, Daphne *Vanishing Cornwall* Victor Gollancz 1967
  Rebecca
Evelyn, John *The Life of Mrs Godolphin* Ed. 1847
Fiennes, Celia *The Journeys of Celia Fiennes* (ed. C. Morris) 1947
Fisher, John *The Origins of Garden Plants* Constable 1982
GIlbert, Davies *The Parochial History of Cornwall* 1838
Girouard, Mark *Life in the English Country House: A Social and Architectural History* Yale University Press 1978
Gore, Alan and Fleming, Laurence *The English Garden* Michael Jospeh 1979
Grigson, Geoffrey *Freedom of the Parish* Anthony Mott 1982
Halliday, F. E. *Richard Carew of Antony* Melrose 1953
  A History of Cornwall Duckworth 1959
Hardy, Emma *Some Recollections* (ed. Evelyn Hardy and Robert Gittings) Oxford University Press 1979
Harrison, E. *Gunners, Game and Gardens* Leo Cooper 1978
Hawker, R. S. *Footprints of Former Men* 1870
Henderson, C. *Essays in Cornish History* Clarendon Press, Oxford 1935
Jenkin, A. K. Hamilton *The Cornish Miner* Allen & Unwin 1927
Kingsley, Charles *Westward Ho!* Dent 1976
Journals of the Royal Institution of Cornwall (various)
Lake, W. *Parochial History of Cornwall* Truro 1868
Leland, J. *Itinerary* (ed L. T. Smith) 1906
Lysons *Magna Britannica*
Norden, John *Speculi Britanniae: Cornwall* 1728
Pevsner, N. *The Buildings of England:* Cornwall 1951
Polwhele, R. *The History of Cornwall* 1803
Redding, C. *Illustrated Itinerary of the County of Cornwall* 1852
Rowse, A. L. *Sir Richard Grenville* Jonathan Cape 1937
  Tudor England Jonathan Cape 1941
  The Byrons and Trevanions Weidenfeld & Nicholson 1978
Tangye, Michael *Tehidy and the Bassets* Truran 1984
Tolstoy, Nicholas *Half Mad Lord, Thomas Pitt 2nd Baron Camelford* Jonathan Cape 1978
Twycross, Edward *Mansions of England and Wales: County of Cornwall* 1846
Vyvyan, J. L. *The Visitations of Cornwall* Exeter: Wm. Pollard & Co., 1887
Vyvyan, Clara C. *The Old Place* Museum Press 1952
  Helford River Peter Owen 1963

p6   The *Mary Rose* was the second largest ship in Henry VIII's fleet that assembled at Portsmouth in 1545. When French ships sailed suddenly into view the English fleet, taken by surprise, prepared to attack. As the *Mary Rose* went into action she began to heel over to port. The sea poured into her open lower gunports and she sank immediately. Roger Grenville and her entire crew were drowned.

Now the *Mary Rose* has been brought to the surface and her contents closely examined. Did the pewter wine flagon, the candlesticks and gold coins found in the officers' quarters aft belong to her gallant captain? And did he use the small wooden gaming board and the tiny pocket sundial made of boxwood that have survived more than four hundred years under the sea? These tangible reminders of life on board make Roger's death more poignant.

The age of Elizabeth I seemed to suit the energetic, restless Grenville temperament and test its mettle. It was a time when men could fight or explore the world in the service of their country, a time when heroism and honour had meaning.

Sir Richard Grenville is Kingsley's hero in *Westward Ho!*, where there is a description of him strolling in the garden of old Stowe with his young godson Amyas Leigh, to whom he gives fatherly advice on the dangers and rewards of falling in love:

> At one turn they could catch, over the western walls, a glimpse of the blue ocean flecked with passing sails; and at the next, spread far below them, range on range of fertile park, stately avenue, yellow autumn woodland, and purple heather moors. . . A yellow eastern haze hung soft over park, and wood, and more. . . And close at home, upon the terrace before the house, amid romping spaniels and golden-haired children, sat Lady Grenville herself, the beautiful St Leger of Annery, the central jewel of all that glorious place, and looked down at her noble children, and then up at her more noble husband, and round at that broad paradise of the West, till life seemed too full of happiness, and heaven of light.

As sheriff and loyal servant to the Queen, Sir Richard made it his business to expose and arrest Cornish Catholics and to keep a sharp look-out along the coast for boats landing Jesuit priests from Douai or Rome. In the 1570s fierce legislation was introduced to counteract the papal bull of excommunication which ordered all Catholics to withdraw their allegiance to Elizabeth I. It became an act of treason to celebrate mass, and anyone caught doing so was severely punished. Staunch Catholics were driven underground and many were ready to die for their faith. Sir Richard Grenville was feared and hated by these brave men, for he showed them no mercy.

Sir Richard is chiefly remembered as a seafaring man, a friend of Sir Walter Raleigh and rival of Sir Francis Drake. Appointed by the government to supervise the defences of Cornwall upon the outbreak of war with Spain in 1585, Sir Richard was chosen to chase off any Spanish ships found lurking in the Irish Channel after the Armada had been sighted off the English coast in July 1588. It was a hard and dangerous task, for violent storms swept the Irish coast that autumn and twenty of Spain's finest galleons were shipwrecked.

When Drake's expedition to Lisbon failed, Sir Richard seized his chance and in 1595 set off as vice-admiral on the *Revenge* to capture the Spanish *flotas* or treasure fleets which passed through the Azores from the Indies en route for Spain. The English ships patrolled the seas for four months waiting for the *flota* to arrive. By then sickness was rife. While they were in the midst of disinfecting the ships with vinegar and bringing clean ballast on board, Spanish ships from the Armada appeared out of the blue and attacked. Raleigh's report indicates that Sir Richard refused to turn tail, which would have been to 'dishonour his self, his country and her Majesty's ship', and faced the enemy bravely. The fighting was heavy. Sir Richard remained on the upper deck until midnight, when he received a severe head wound. By morning the noble *Revenge*'s masts were torn and the powder had virtually run out. Sir Richard was all for sinking the ship with what remained, 'that nearby nothing might remain of victory or glory to the Spaniards', but the captain and master refused to let him and he was taken aboard the Spanish flagship. He died a day or two afterwards.

Sir Richard's grandson, Sir Bevil Grenville, was equally courageous. He succeeded to Stowe in 1636, a few years before the Civil War, and was the first to raise King Charles's standard in the West and call his tenants to arms. Under Sir Ralph Hopton, he led the Cornish forces to victory at Braddock Down in January 1643, and followed this with another success at Stratton, a few miles to the south of his beloved Stowe. The fact that he knew every hill, wood, and dell so intimately proved invaluable as he orchestrated the battle and rounded up over a thousand prisoners. Two months later this gallant soldier was mortally wounded at Lansdowne near Bath, together with fellow Cornishmen Godolphin and Trevanion. This was the turning point of the Civil War in the West, for with the death of these charismatic leaders the fighting spirit and will of the Cornish soldiers weakened.

p 6★★ John Grenville defended the Scilly Isles until their final surrender to Cromwell in 1651 and then went into exile. He was allowed to return to Stowe in 1659, mistakenly trusted by the Parliamentarians, for he began at once to plot their overthrow. He had appointed a relation of his, Nicholas Monck, as rector of Kilkhampton and immediately sent him to Scotland to persuade his brother, General George Monck, to help restore Charles II to the throne. Monck and Grenville met secretly, exchanging messages from the King, and on 1st May, 1660, Grenville delivered a letter from Charles to both the Commons and the Lords. The Restoration was under way.

p12 Wrecks were a common occurrence on this savage north coast. Between the years 1824 and 1874 as many as eighty vessels were swept on to the rocks round Bude. The *Caledonia*, with a crew of ten, was wrecked off Sharp's Nose, a headland just south of Morwenstow, in 1842. Hawker helped carry the dead up the cliff 'to await, in a room at my vicarage which I allotted them, the inquest', and cared for the sole survivor, Edward Le Dain of Jersey. The figurehead of the *Caledonia* – a white-painted wooden figure brandishing sword and shield – still rests over the captain's grave in Morwenstow churchyard.

The ghosts of drowned sailors whose bodies were laid out in the lychgate house for burial haunted Hawker, and yet he was drawn to the sea. The hut he constructed from the timbers of wrecked ships still stands on the cliff. When the weather was clement he would sit here in his claret-coloured coat, fisherman's jersey, and long sea-boots and meditate on the paradoxes of the world about him. Here too, he would read aloud to his wife and compose his poetry, much of it filled with images of the sea.

p22   In 1483 Richard Edgcumbe, together with other prominent West-countrymen, joined the Duke of Buckingham in an attempt to overthrow Richard III who, it was rumoured, had murdered the two sons of Edward IV in the Tower of London. The revolt failed. Savage storms prevented Buckingham from crossing the Severn on his way to Exeter and he was caught and executed without trial at Salisbury. The same storms made it impossible for Henry Tudor to land and he was forced to return to Brittany. All those implicated in the plot were outlawed. Edgcumbe took refuge at Cotehele where he was eventually tracked down by Richard III's dreaded local agent, Sir Henry Bodrugan. Miraculously he managed to slip through the cordon, and fled towards the river by a steep path that is today part of the Cotehele garden. He saved himself by filling his cap with stones and flinging it into the water so that Bodrugan's henchmen, in hot pursuit, thought he had jumped in and drowned. When his pursuers had gone Richard sailed quietly away to Brittany to join his King.

Two years later, Henry Tudor landed at Milford Haven and won his crown at Bosworth Field. Richard Edgcumbe was knighted after the battle and made Comptroller of the Royal Household, Chamberlain of the Exchequer and Member of the Privy Council. At the same time he acquired the estates of Sir Henry Bodrugan near Mevagissey.

p31   Richard was a man of letters. Virgil, Homer, Pliny, Holinshed, Chaucer, Spenser, Marlowe, and Shakespeare were his companions and he taught himself Greek, Dutch, French, Spanish, and Italian. He so enjoyed 'sweet relished phrases' of Italian that he set about translating a long poem by Tasso which was published in 1591.

His fame rests above all on his *Survey of Cornwall* (1602), an account of the topographical, political, social, and economic structure of the county. He is a shrewd observer, witty and outspoken. Crammed with interesting facts, the *Survey* is also full of amusing anecdotes which tell us much about the workings of an Elizabethan gentleman's mind. His duties as justice of the peace, deputy-lieutenant of the militia, and sheriff of Cornwall took him all over the county at a critical time in English history. A Spanish invasion was a constant threat. There was panic at Antony when news came that the 'Spanish floating Babel', to quote Carew, was nearing Plymouth. His friendship with other members of the gentry involved in public service kept Carew up-to-date on all the local gossip and gave him first-hand knowledge of matters otherwise outside his ken. Sir Francis Godolphin, for example, supplied him with the information on mining that he needed for his *Survey*.

p31★★'The fish thus taken are commonly bass, mullet, gilthead, whiting, smelts, fluke, plaice and sole,' says Carew, who, like any keen fisherman, is proud of what he has caught. 'The pond also breedeth crabs, eels, shrimps, and [in the beginning] oysters grew upon boughs of trees which were cast thither to serve as a hover for the fish.' Carew would watch these trapped fish for hours, noting their reactions to the food thrown to them and their different swimming patterns according to the time of year.

He even planned, at one stage, to build 'a little wooden banqueting house on the island in my pond'. In Elizabethan times small banqueting houses, of fanciful architecture to match the exotic food served in them, were often built either in a turret on the roof or in the garden. After the main meal was over, the company would retire to these banqueting houses to be served the void (in mediæval times eaten standing as the tables were 'voided') or dessert of spiced delicacies and sweet wine. Carew's banqueting house was to have had a square ground plan, with a round room downstairs and a round turret containing a square room above. A tiny kitchen, buttery, and storehouse for fishing rods led off these two rooms, as well as space for 'cupboards and boxes, for keeping other necessary utensils towards these fishing feasts'. We do not know why such a delightful project was abandoned.

p36   He also raised its status when he was knighted in 1618 and made Vice-Admiral of Devon. Unfortunately for him in the course of his duty he arrested a pirate who, as it turned out, had a friend at Court, and it was Eliot who found himself imprisoned. Disillusioned with such corrupt ways he fervently opposed the unconstitutional religious and financial measures of Charles I. In 1629 he refused to pay the King's forced loan and persisted in defending the liberties of the Constitution. For his moral courage he spent the last years of his life in the Tower and died of consumption in 1632. A few days before his death an unknown artist painted him in a lace-embroidered nightshirt, looking resolute and unrepentant. This poignant portrait hangs in Port Eliot today.

p43   Mystery surrounds Arthur's birth – some say he was cast from a shining ship and borne on a mighty wave to Merlin's feet one dark and stormy night. One of the more popular legends tells that Uther Pendragon, King of All England, fell in love with Ygerne, the beautiful wife of Gorlois, Duke of Tintagel. Uther invited Gorlois and his wife to his castle in Carlisle and, during a lavish entertainment, told Ygerne of his love for her. The lovely Ygerne rejected the King, and she and Gorlois fled back to Tintagel to prepare for a siege. Because of its inaccessible position, Tintagel could be defended by relatively few men, so Gorlois left Ygerne and went to defend his other castle, Terrabil. Here he was killed in a bloody combat with Uther's army, which then went on to overpower the handful of men at Tintagel. Uther forced the unhappy Ygerne to marry him and a son, Arthur, was born.

The babe was handed over to Merlin secretly in a cave on the left of the postern gate of the castle and brought up in his care. When King Uther was dying, the old wizard summoned all the barons to his bedside so that they could hear the King proclaim Arthur his heir. When the barons still fought for the throne, Merlin persuaded the Archbishop of Canterbury to call all the nobles as Malory says, to 'the grettest chirch of London' where, one by one, each had to try to pull a sword out of a steel anvil set in the middle of a marble slab. On the hilt of the sword was written: 'Who so pulleth out this

sword of this stone and this anvil is likewise King, born of all England.'
Only Arthur succeeded, and he was duly crowned.

Legend has it that the Round Table of King Arthur is buried beneath a
circular barrow not far from the main road between Tintagel and
Boscastle. This mound was once the site of Bossiney Castle, used as an
outpost for Tintagel Castle and recorded in Domesday Book as being held
by Count Robert of Mortain.

Tintagel is also associated with the Tristan legend. The twelfth-century
poet Beroul, in his *Roman de Tristan*, one of the earliest known versions of
the legend, makes Tintagel the residence of King Mark, uncle of Tristan
and husband of Iseut or Isolde. The beginning and end of Beroul's poem
have not been preserved, but scholars have reconstructed the missing
episodes. In this version Tristan became the lover of King Mark's wife.

Iseut then avoided discovery by lying on many occasions but was finally
caught *in flagrante dilecto* and sentenced to death. They escaped, however,
and after further adventures were finally united. Wagner's opera *Tristan
and Isolde* gives a nobler version of the story, and Matthew Arnold's poem
'Tristram and Iseult' is beautiful and uplifting.

Another poem describes how their bodies were brought back to
Cornwall and buried in the church of Tintagel, one on each side of the
nave. The story is told of two trees that grew miraculously one from each
tomb, their branches meeting in an embrace over the apse. Three times
King Mark had the trees cut down, and three times they grew again.

p46   Bishop Grandisson of Exeter, in his Register for 1357, records how,
on December 27th, 1356, a host of armed men 'furiously entered the
chancel and with swords and staves cut down William Penfoun, clerk', and
desecrated the church with human blood. The bishop gives no reason for
the crime, so we are left to wonder whether William was murdered for
corrupt or immoral practices or whether he was set upon by thugs.

The family fared better in the fifteenth century, though not without
some narrow squeaks. Documents in the Public Record Office reveal that
in 1460 John Wynard of Plymswood Farm (which still stands) 'broke open
two of John Penfound's houses and carried away his goods'. He also tried
to kill his neighbour, and 'on the following Sunday went with one hundred
men to his house' in an attempt to lynch him. What on earth caused such
hatred between the two men?

The following year, in 1461, John Penfound was granted the Office of
Weigher of tin and lead in Devon and Cornwall for life. A few years later
he was granted 'Office of Keeper of the Parks of the King at Hellesbury
[now Michaelstowe] and Lanteglos', with a salary of £4 11s 3d. But in 1476
he appears to have gone to London on some wild extravagant escapade
with a John Beaumond and was afterwards outlawed 'for failing to appear
in the Court of Common Pleas to answer Nicholas Mille, citizen and tailor
of London, touching a debt of £150 £8s 7d' - a considerable sum of money.

By the sixteenth century the Penfounds seemed bent, by fair means or
foul, on acquiring more land. A petition to the Lord Chancellor in 1533
accuses Thomas, William, and Edmund Penfound of having unlawfully
seized '300 acres of pasture, 300 acres of land, 100 acres of meadow, 40
acres of wood, and 600 acres of furze and heath in Cornwall'. A similar

complaint was lodged in 1544 by Crystofer Coke who maintained that Burracott farm (which still exists at the bottom of the hill) belonged to him by right, but that 'Thomas and Edmund Penfound obtained some of the title deeds, forged others and claim the lands'.

p51   The Mohuns had lived at Hall and Bodinnick, across the water from Fowey, since Edward III's time, and were a respected Cornish family. Sir John Mohun, son of the 1st baronet, was created a peer in 1628 and became Lord Mohun of Okehampton. The title became extinct on the death of his great-grandson Charles, the 5th and last Lord Mohun. Charles had a quick and vindictive temper. He was twice tried and acquitted of murder but his luck ran out in 1712, when he challenged the Duke of Hamilton to a duel in Hyde Park. The two men disliked each other intensely, owing to a disagreement over money, and were resolved to fight to the death. Jonathan Swift mentions the affair in his *Journal to Stella*. Both men seem to have given a mortal thrust at the same instant and both died in a pool of blood. 'While the Duke was over him,' says Swift, 'Mohun shortened his sword and stabbed him in the shoulder to the heart.'

p53   During the last few months of 1642 the Royalist commander Sir Ralph Hopton had organised an army of volunteers, consisting in the main of tenants and servants of the Cornish gentry. This small band, only 1,500 foot with a few guns, was a formidable force prepared to fight with feudal devotion. By November Hopton felt confident enough to advance on Exeter, but the expedition was a failure due to inadequate provisions and insufficient discipline, and his men had to retreat to Bodmin. In January 1643 the Parliamentary forces retaliated and crossed the Tamar, the Scottish Colonel Ruthven at their head. Fortunately for Hopton, Ruthven did not wait for the main body under the Earl of Stamford but advanced from Liskeard and drew up his troops on the downs in front of the tiny church of Braddock, not two miles from Boconnoc woods where Hopton's men were encamped.
    The battle was soon over. After prayers had been said the brave Royalists advanced, the foot in the centre, the cavalry on the outside. Sir Bevil Grenville suddenly ordered the foot to charge and they did so with such ferocity that the enemy turned and fled. 'But when resistance was over,' wrote Clarendon, 'the Cornish soldiers were very sparing of shedding blood, having a noble and Christian sense of the lives of their brethren.' This first battle on Cornish soil was a resounding victory for the Royalists, whose losses were slight. Twelve hundred Parliamentarians and five guns had been captured, and Hopton wasted no time in attacking and capturing Saltash four days later.
    After more than a year of bitter fighting beyond the Tamar, the Parliamentarians returned to Cornwall at the end of July 1644. Their new leader, the Earl of Essex, grossly underestimated the hostility and hatred of the Cornish peasantry, but when King Charles set up headquarters at Boconnoc Essex realised he was done for. The King's men were joined by Grenville's loyal supporters, who had taken Bodmin and were camped at Lanhydrock, while Hopton's army provided more reinforcement. Essex was virtually encircled, his 10,000 men confined to an area of a few square

miles between Lostwithiel and the sea.

His only hope was to occupy Fowey and pray for the arrival of more troops by sea, but he stupidly allowed the King to occupy the east bank of Fowey River, thereby losing control of Fowey itself. By the end of August, after a month of ransacking the houses in the area for food, Essex's men decided to make a dash for it. On the dark, misty night of August 30, the bedraggled foot made their way through the mud to Fowey. They got as far as Castle Dore, some two miles to the north of the port, where they were attacked by the Royalists the next morning. Essex slipped away and sailed for Plymouth in a fishing boat, leaving Skipton, who had brought up the rear with the artillery, to surrender. The terms were honourable but harsh. The wounded were to be sent by sea from Fowey to Plymouth, while the rest of the army was to march out of Cornwall under guard. The Parliamentarians had to suffer not only the abuse of the Royalist army but also the harsh scorn of the Cornish peasants, who stripped them of their clothing and beat them. Only 1,000 out of the 6,000 men who left Lostwithiel reached the safety of Poole in Dorset, for brutality, starvation and exposure took their toll, while the Parliamentarian horse, riding through the night, evaded the Royalists and escaped to Devon. In time this escape was seriously to undermine Charles's victory.

Meanwhile, King Charles and his supporters, who had been showered with honours for their services, feasted at Boconnoc.

p53★★   Thomas Pitt was born at Boconnoc in 1775. Separated from his parents, who spent most of their time in London or abroad, Thomas was left in the care of a tutor. Deprived of playmates and human affection, yet conscious of his privileged position as heir to a great estate, he grew to be emotionally vulnerable yet proud. His father died in 1793, soon after Thomas had joined the navy, and his career henceforth was both turbulent and tragic. After his death Boconnoc passed to his sister Anne, who had married William Wyndham, Lord Grenville. He died in 1834, when George Matthew Fortescue inherited it.

p54   Ever since John Glynn had replaced Thomas Clemens as deputy-steward of the Duchy, he had been a doomed man. Clemens bitterly resented his appointment and was determined to make him pay for it with his life. In January 1469 his henchmen beat up Glynn and his servants while he was holding the court of the manor, tore up the court rolls and held him under duress until he signed a £200 bond agreeing not to prosecute his assailants.

The following year, Clemens and his men raided Morval and stripped it of furniture and provisions. John's wife alleged that they stole feather bedding, pillows, tapestries, cushions, silver plate, and 400 gallons of ale. Not content with this, they drove off 14 oxen, 10 kine, 60 bullocks, 8 horses, 400 sheep, and 10 sows from the surrounding pastures. Although the Glynns issued a writ of appeal of robbery against him, Clemens continued to roam freely and a few months later struck the final, fatal blow. His spies informed him that John intended to go to Tavistock fair. An ambush was planned, and at 4 a.m. they pounced on him at Higher Wringworthy and hacked him to pieces. His grieving widow describes these events in a petition to Parliament complaining of the lawlessness of Cornwall. His murderers 'clove his head in four parties and gave him ten dede wounds in his body; and when he was dede, they kutt off one of his legges and one of his armes, and his hede from his body'. They also stole his purse, signets of silver, sword, dagger, and even his cloak. Clemens harboured the assailants until they were implicated and then went into hiding himself, but it is not known whether he was ever caught and hanged.

p55   The main residence of the Trelawnys had been the manor of Trelawney (from which they took their name) in the parish of Altarnun near Launceston. This passed into other hands when the male line of the elder branch of the family died out in the mid fifteenth century on the death of Richard Trelawny. Richard's younger brother John continued the line and his descendant Sir Jonathan, the first of a long line of baronets, settled at Trelawne early in the seventeenth century.

p55**   Bishop Trelawny took holy orders in 1673 and became Bishop of Bristol in 1685. A staunch Royalist like his father before him, he supported James II during the Monmouth Rebellion and remained loyal to the Stuart cause until the King's catastrophic religious policy brought about an irrevocable rift.

In 1687 James II published his Declaration of Indulgence exempting Catholics and Dissenters from penal statutes and, by the use of his dispensing power, introduced both Dissenters and Catholics into all departments of state. The following year he went further and issued a proclamation to force the clergy to read the detested Declaration from their pulpits. This was the last straw as far as Bishop Trelawny was concerned. He and six other bishops refused, and were immediately imprisoned in the Tower. There was uproar in Cornwall while the trial went on, and public rejoicing in the streets when the seven bishops were acquitted and released. Hawker's poem is thought to refer to this incident.

A few weeks later, invited by Protestant nobles enraged at James's policies, William of Orange landed in England. Trelawny swore an oath of allegiance and was rewarded with the bishopric of Exeter and later that of Winchester. He died in 1721 at his residence in Chelsea but his body was brought back to Cornwall and buried at Pelynt, close to the cherished grey stone house of Trelawne.

p55 †   The Trelawnys controlled not only fishing rights but the right to unload sand, which with seaweed supplemented the farmyard manure on Cornish farms. Barges fitted with heavy canvas bags kept open by iron hoops dredged the sand from the sea-bottom off Looe or Talland Bay and then proceeded up-river to deposit their load at Trelawne Moor.

As the demand for fertiliser increased, so lime-kilns appeared on the river bank, limestone quarried outside Plymouth being the nearest supply. Behind the kilns between Shallow Pool and Watergate, the nearest point on the river to Trelawne, traces remain of the deeply rutted tracks made by convoys of packhorses as they trod through the Trelawne woods with as much as four hundred pounds of lime straddled in bags across their saddles.

Revenues from the silver and lead mines higher up the valley, under the hills to the north and south of Herodsfoot, were another source of Trelawny income, although no vast fortune was made as was the case with the tin and copper mines further to the west. Shafts were sunk down to the ore-ground in the 1840s, but miners had searched for scraps of shimmering metal for over a century before and continued to do so after the mines were abandoned at the end of the nineteenth century.

In the seventeenth and early eighteenth century the manufacture of West Country serge brought considerable wealth to Devon merchants, and evidence of fulling-mills on the Trelawne estate and elsewhere in the Looe river valley shows that this eastern part of Cornwall shared in the trade. The Pelynt weavers would bring their woollen cloth to the fuller, who lived in a cottage of cob behind Trelawne Mill. The fuller's job was to scour, dry and stretch the cloth on racks and then to full or thump it. He would first of all fold the cloth, rubbing it with soap between the folds, and then place it beneath a heavy wooden bar known as the perch, which would hammer the cloth as the water-wheel turned.

p58 Carnsew was also keenly interested in mining and welcomed Sir Francis Godolphin as a visitor. Unlike Sir Francis, William had no valuable tin mines on his land, but he dabbled in various mining ventures nonetheless. A London company, providing financial backing to prospect for copper, lead and possibly silver, relied on Carnsew for honest on-the-spot intelligence and advice. They sent down a young and rather self-opinionated German mining engineer named Ulrich Frose to oversee work at the Treworthy mine on the coast near Perranporth. Ulrich stayed at Bokelly over Christmas, but the realisation that there was not enough capital to dig deeper made for a somewhat strained atmosphere. The two men got on each other's nerves and Ulrich was no doubt miserable in the cold and damp winter climate. Carnsew never grew rich through his mining activities, but the excitement they engendered outweighed any disappointment.

p64 John Molesworth was the Queen's Commissioner for the Duchy of Cornwall and married Catherine, daughter and heiress of John Hender of Botreaux Castle, near Tintagel. His grandson settled in Jamaica and eventually became Governor of the colony. William III made him a baronet in 1688, as a reward for loyalty during the religious persecutions which took place in Jamaica during James II's reign. On his death in 1689 the baronetcy and his Jamaican fortune passed to John's elder brother, Sir John Molesworth of Pencarrow.

p66 High-spirited Emma and her brother used to go fishing for trout in the stream at the bottom of the garden, 'puss always on a little boulder in the middle of the stream – at intervals putting her paws right into the water'. This might seem a childish pursuit for a girl of twenty, but Emma's parents, over-conscious of their former gentility, kept to themselves in their now impoverished state and only visited two families in the entire neighbourhood. Emma thus played with her brothers and kept the one remaining servant, Anne Chappel, company. Anne 'kept a basket of crockery under a tree and a great kitchen table, and we had pleasant reading and working hours and many meals there'.

Modern-day Bodmin does not intrude, and one can imagine her sister's wedding day, as Emma described it. Kirland would have looked its smartest and gayest as the guests clustered round the entrance porch to wave goodbye to the bride and groom. Emma left with them in the dog-cart, for she was to housekeep for her sister. Helen Catherine had married the rector of St Juliot, a remote village on the wild north coast, near Tintagel, and it was here that Emma met Thomas Hardy, a young architect sent to restore the ruined church. Canny old Anne Chappel had prophesied years before that Emma would marry a writer and Helena vicar. As she left Kirland destiny was taking Emma one step nearer the fulfilment of her romantic dreams.

p68   Prior Vyvyan was a powerful figure in his day, although the whole ethos of monastic life was changing and the old order was soon to disappear. Lay officials were by now managing the estates belonging to Bodmin Priory–Nicholas Prideaux was Prior Vyvyan's steward or man of affairs – and they were not slow to turn such a position to their advantage in the years before the Dissolution and afterwards. Prior Vyvyan died in 1533 and was succeeded, after much politicking, by Thomas Mundy, canon of Merton Abbey in Surrey. Although Bodmin was far away from London, it was only a matter of time before Henry VIII's grasping agents arrived to strip the priory bare and deprive him of his wealth and privilege. Mundy acted in the nick of time. At midsummer in 1537, he summoned his brethren because 'he did here that the King's Majesty would take his pleasure upon their house, and, therefore, he thought it good to give unto such as had beene good to the house, some leases or other preferments, to the interest they should be the better to them hereafter'.

Mundy's suggestion was even less altruistic than it appears. Those whom he felt should receive 'leases or other preferments' turned out to be his nephew and niece and Nicholas Prideaux. Like Prideaux, Mundy was a bachelor, and the two men planned to tie the families together to the mutual advantage of both. Mundy granted a ninety-nine-year lease on Rialton to his brother John at £60 per year, a very reduced rent. This was a clever move, because the normal lease for a Crown property was only twenty-one years and Henry VIII soon made longer leases illegal because they deprived the Crown of substantial revenues. John Mundy's son William married Nicholas Prideaux's niece Elizabeth, and her brother William Prideaux married John Mundy's daughter Joan, thus securing the Rialton property to both families.

On her marriage to William Prideaux, Joan Mundy was granted a ninety-nine-year lease of the manor, customs, fisheries and advowson of Padstow at an annual rent of £10 7s 8d. In 1544 the freehold of the manor was sold by the Crown to a Londoner, John Pope, for £1551, perhaps in lieu of money owed.

Pope sold it immediately to Nicholas Prideaux. The expense of war in the reigns of Henry VIII, Mary, and Elizabeth, and the Crown's constant need of ready cash, made the sale of these erstwhile ecclesiastical estates inevitable. By the end of the century almost all had been sold. It was, after all, a unique opportunity for the gentry to acquire more land and hence to consolidate their power.

p71   Friends of Emma's, Captain and Mrs Serjeant, were living at St Benet's in the 1870s. 'A chapel was within, large, full of lumber, a fine large hall and grandly wide and windowed principal staircase with a spiral stair, all in good repair and very habitable. Nevertheless it was a dark, gloomy and not very healthy abode . . .' Emma seems to have made the most of the library during her stay, though she remarks that no one read the 'much mildewed' books save her.

One wet day I rubbed them dry and enjoyed a good deal of browsing for some days amongst them. There were sloping lawns, terraced gardens, and a fishpond, but everywhere an old-time sadness prevailed. One's soul was not refreshed out of doors, where we were amongst the bones of the former immured people, though no cemetery could be seen or known.

The atmosphere was rendered even gloomier by the doleful presence of the rector who actually owned the place and who was looked after by the Serjeants. He was an invalid and recluse whom Emma never saw, for he remained shut in his room.

Emma does recall one exciting incident which would have temporarily relieved the gloom. Lanivet was

in the heart of the China Clay works, which made the country walks hideous with yellow mud, Stamping Plant always working, and pools and cisterns of white and of green water, and turned-up arid soil – but my friends were enriched by the finding of clay on the estate whilst I was there. A large bowl was brought to us all to look at one evening and pounded and prodded and handled lovingly and chuckled over; it proved to be of excellent quality, whitest of white and pliable and the agent came and settled it all. It was a valuable asset to their income, but if it had all been *under* the house what would they have done?

p72   After the Prayer Book Rebellion of 1549 and the swift execution of the rebel leaders (one of whom was Humphry Arundell of Helland, a cousin of the Lanherne branch), most of the Cornish gentry accepted the introduction of the new Protestant Ordinal and adapted themselves to the religion required of them by the monarchy. Not so the Arundells, who continued to worship as devout Catholics despite severe persecution. But it was not only their recusancy that set them apart from the other gentry. The Arundells were socially grander than most other Cornish families and moved in elevated circles outside the county.

The 12th Sir John's brother Thomas, who began his training as a lawyer in the household of Sir Thomas More and Wolsey, had married Queen Catherine Howard's sister and bought Wardour Castle in 1547. The 13th Sir John married the daughter of the Earl of Derby, the widowed Lady Stourton, and succeeded at Lanherne. Their aristocratic connections probably affected their relationship with the more modest landed gentry of the county, who either held them in awe or envied their position.

Lanherne was a haven for all Cornish Catholics as persecution gradually forced them underground or abroad. The Arundells harboured many a seminary priest in their household, Catholics coming from far and wide to celebrate mass in secret at Lanherne. For his continued support of the

Catholic faith even Sir John, with all his influence at court, was sent to the Tower in 1584, and was kept in semi-custody until his death at Isleworth, near London. His widow retired with her priest to the Arundell estate at Chideock in Dorset, but it was only a matter of time before spies informed the Queen's servants of her whereabouts. The house was searched in 1594 and evidence found to support the suspicion that daily mass was being said, a punishable offence under the Act of 1571. The faithful chaplain was executed, and Lady Arundell retreated to a convent in Brussels.

p76   Edmund Tremayne, the builder of Collacombe, acted as a kind of West Country agent of the Queen and was well-known at Court. The 4th Sir John may have met him through his father, who spent his life in royal service, and, being pleased with the excellent work of his own plasterer at Trerice, recommended him to Tremayne. The dates on the overmantels support such a theory: 1572 and 1573 at Trerice and 1574 at Collacombe. He may then have gone to Buckland Abbey, the great house of Sir Richard Grenville, for the overmantel in the hall there bears the date 1576 and is very similar in appearance.

After the death of his first wife Katherine, Sir John married Gertrude Dennys of Holcombe and their son, an ardent royalist, became the 5th Sir John Arundell.

He was born a few years after the completion of the new house and Sir John spent most of his life as an M.P., first for the little borough of Mitchell and then for the County. He was sixty-five when the Civil War broke out, but was such an ardent supporter of the King that he was appointed governor of Pendennis Castle after the death of Sir Nicholas Slanning at the siege of Bristol in 1643. Here he harboured poor Queen Henrietta Maria on her flight to France, and later Prince Charles, who escaped to the Scilly Islands only a few weeks before Fairfax demanded the garrison's surrender. The seventy-year old Sir John despatched his reply immediately. 'I resolve that I will here bury myself before I deliver up this Castle to such as fight against his Majesty, and that nothing you can threaten is formidable to me in respect of the loss of loyalty and conscience.' The brave old man held out for five months despite appalling hardships. Plague broke out and many of the inmates were reduced to starvation. Finally, he wrote to the Prince saying that he could not allow women and children to die of hunger and agreed to Fairfax's terms.

His second son Richard succeeded to a much impoverished Trerice, for his elder brother had been killed at Plymouth in 1643. His father did not live to see the return of the monarchy but soon after the Restoration Charles II remembered his loyalty and that of his two sons by creating Richard Baron Arundell of Trerice. One hundred years later, on the death of the 4th Baron Arundell in 1768, the estate, like that of Ebbingford, passed to his wife's nephew William Wentworth.

p77   John Robartes was a Parliamentarian, and for a few momentous months in 1644 Lanhydrock was the headquarters of the Parliamentary forces. The 2nd Lord Robartes fought bravely at Edgehill and Newbury and was promoted Field Marshall under Essex. That August his house was seized and occupied by the Royalists and his estates assigned by the King to

Sir Richard Grenville. Robartes himself managed to escape by sea to Plymouth, but his children were detained as prisoners. He regained Lanhydrock under the Commonwealth, devoting much of his time thereafter to developing the estate. But he came to disagree violently with Cromwell's decision to rule the country without Parliament and, after Cromwell's death, he played an important role in securing the return of Charles II, for which he was rewarded at the Restoration.

p84   Times were uncertain during the mid-sixteenth century; years of bitter warfare between Emperor Charles V and François I of France were interrupted by truces no sooner made than violated. Wolsey's policy of courting first one power, then the other, antagonised both. The King alienated himself further by his attempts to divorce Catherine of Aragon, Charles V's aunt, and by his final breach with Rome. François I had no reason to like him either, for England had mounted expeditions to France and was not to be trusted. When the Emperor Charles V and François I resumed hostilities in 1542, Henry joined in on the side of the former. An English army landed on French soil and conquered Boulogne, a foolish undertaking since it achieved nothing militarily speaking but increased the likelihood of a French attack on England.

p86   John Treffry is said to have captured the French royal standard at the battle of Crecy in 1346. He was rewarded with a knighthood and the right to quarter arms with the French *fleur de lis*. His descendants continued to advance in royal service. Another John Treffry went into exile with Henry Tudor, landed with him at Milford Haven where he was knighted and fought at the battle of Bosworth in 1485. The family then prospered under Henry VII during whose reign much of Place was rebuilt. John's brother William who succeeded him at Place became Usher of the King's Chamber, Controller of the coinage of tin and Surveyor of Customs within London.

p86★★   Thomas was right to fortify his home. A century later the threat of attack from Spain caused John Treffry to assemble pikes, halberds and other weapons at Place. By 1568 Anglo-Spanish relations were strained to breaking point. When Spain sent the ruthless Duke of Alva to crush the Protestant resistance in the Netherlands, England retaliated. Privateers drove two Spanish ships into Fowey harbour. The gold coins on board were taken to Place to be weighed and counted in full view of the townsfolk before being taken to the Tower of London. This treasure, a huge loan from Genoese bankers to the Spanish king, was on its way north to pay Alva's hungry and discontented troops. The Spanish government promptly placed an embargo on English shipping and England counter-attacked appropriating any Spanish ship in an English port. The treasure was not returned, and Alva's troops mutinied.

p89   Tantalisingly, although a complete set of architect's drawings have been preserved, they are unsigned and have no precise dates. Nor were many of the designs followed. Records exist, however, of payments to a Mr Edwards, and it is likely that he was Thomas Edwards of Greenwich and the architect of Trewithen. Very little is known about Edwards, and no record has been found of his having designed houses elsewhere in England. Enough evidence exists however to show that he managed to build up a considerable practice in Cornwall, and he will be mentioned in subsequent chapters in connection with other houses. His will indicates that he was a mine and ship owner as well as an architect, and these interests no doubt kept him in Cornwall. His association with the mining magnate William Lemon, for whom he designed Princes House in Princes Street, Truro in 1737 and Carclew in 1749, would have led not only to further architectural commissions but to exciting commercial ventures. One can also surmise that Edwards was an admirer or follower of James Gibbs, since he was one of the subscribers to Gibbs' *Book of Architecture* (1728).

p92   He started the great East Wheal Rose lead and silver mine near Newlyn East, one of the richest in the county, and established a reputation as a notorious borough-mongerer. He was sole owner of the pocket boroughs of Tregony, Grampound, and Mitchell, obtaining the latter for Sir Arthur Wellesley, later the Duke of Wellington, in 1806. One of the houses in Mitchell is called Wellesley House to this day.

Sir Kit, as he was known, was a forward-thinking squire who had his tenants' interests much at heart. An early educator in terms of agricultural practice, he laid the first clay-pipe land drains in Cornwall. The patron of Richard Trevithick, he also commissioned the first steam threshing machine for Trewithen. This is now in the Science Museum in London.

p93   By the seventeenth century the gentry of Cornwall were no longer content merely to receive a toll from tinners wanting to work on their land. They were ready to gamble, and formed their own companies, sometimes owning all the shares so that they could win or lose all. As technology improved the shafts could be sunk ever deeper. Rich lodes of copper were discovered beneath the tin, and in the early nineteenth century, when new high pressure engines, smaller and cheaper, were introduced, there was a spectacular increase in production. Cornwall soon became the largest copper-producing area in the world. The population of Cornwall soared as thousands came to find employment in the tin mines, but working conditions were appalling.

p98   In 1485 Henry VII rewarded William Trevanion for his Lancastrian sympathies by making him Esquire to the Body of the King. Henceforward the Trevanions were in royal service. William's son Hugh was one of the commissioners appointed to survey Cornwall's coastal defences in 1539, and was also responsible for collecting funds to pay for war and for arresting pirates. Caerhay's coastal position allowed him to keep an eye on activities at sea between Fowey and Falmouth.

Hugh's descendant Jack Trevanion raised his own volunteer regiment at the outset of the Civil War, as did other leading Cornish families. These plucky infantrymen fought every action from Braddock Down to the siege of Bristol in July 1643, when many were killed. The Cornish foot attacked in three brigades - Sir Nicholas Slanning and Trevanion in the centre, Colonel Buck on the right and Sir Thomas Basset on the left. The charge was a fatal one. Buck was decapitated by a halberd, Basset was hit and Slanning and Trevanion were both shot in the thigh with musket bullets and fell immediately. Jack's bravery was remembered by the King when he knighted his father at Boconnoc the following summer, but nothing could make up for the loss of those gallant Cornishmen who perished at Bristol.

One of Jack's sons, Richard, supported James II in the same selfless way, escorting him to France when his luck had run out. Richard supported his unfortunate monarch when he led an abortive expedition to Ireland in 1690, attending him until his death in 1701.

In the eighteenth century the Byron family enters the Trevanion annals. John Trevanion and the 4th Lord Byron married the two daughters of Lord Berkeley of Stratton, Barbara and Frances respectively. John and Barbara's son William was the last male Trevanion. William's sister Sophia married her cousin, the younger brother of the 5th Lord Byron. This marriage, which took place in the chapel adjoining the gatehouse (now demolished) in 1748, marked the beginning of a series of ill-fated relationships. On William's death in 1767 the estate passed to his other sister Frances and her husband John Bettesworth.

Sophia's son Jack, a handsome rake, was the poet Byron's father, and her daughter Fanny's son was later to marry the poet's half-sister Augusta. Then, two generations later a descendant of John and Frances Bettesworth, Henry, married Augusta's daughter Georgiana. Incestuous passion and depravity were the corollary to these disastrous unions which undoubtedly weakened the Trevanions' moral fibre. But Henry's father, John Bettesworth (he assumed the name of Trevanion in 1801), set the final seal on the family's demise when he decided to pull down the ancient seat of Caerhays, which stood to the north of the present castle. This was an emotional and physical blow from which the Trevanions never recovered.

p102   At the time of the Stuarts, Killigrews had considerable influence at court. Sir William Killigrew, eldest son of Sir Robert Killigrew, was a Gentleman Usher to Charles I and commanded one of the two troops of horse which guarded the King during the Civil War. After the Restoration he became Vice-Chamberlain to the Wueen and a prolific dramatist. His brother, Thomas, also a dramatist, began his royal service as Page of Honour to Charles I. He was Charles II's Resident in Venice during the Commonwealth years and then in 1660 became Groom of the Bedchamber and Master of the Revels. He also managed the Theatre Royal in Drury Lane.

Both brothers and their sister Anne, a Dresser to Queen Henrietta Maria, sat to Van Dyck in 1638. They make a debonair trio. All three have wide-set, intelligent eyes, pointed noses and sensitive fingers. Noble and confident, their fine clothes enhance their bearing. Anne, who points to a rose-bush beside an urn, wears a lustrous gold satin dress with a grey scarf tied across her breast; William, the most serious of the three, is dressed in sober black, while Thomas, who fondles a huge mastiff, wears a breastplate and crimson sash to offset his long golden hair. Like so many of Van Dyck's sitters, they occupy with easy grace a glittering stage, unaware that the play is about to end.

p107   Mary Granville, as she was born, was the great-granddaughter of Sir Bevil Grenville of Stowe, who was killed during the Civil War. Her parents were grand but poor, so Mary was sent to Whitehall as a child to stay with her uncle George who was made Lord Lansdowne in 1711. His wife had been Maid of Honour to Queen Mary and was a friend of Queen Anne. With such connections young Mary would be seen, admired, and eventually wooed, it was hoped, by a rich suitor. The death of Queen Anne in 1714 put an end to these hopes. Lord Lansdowne was imprisoned in the Tower with Lord Oxford for expressing his Jacobite sympathies, and Mary and her parents fled to Buckland in Gloucestershire. Knowing that they would not be welcome at the new Hanoverian Court, and poorer than ever, the question of finding a husband for Mary, now fifteen, loomed large. On his release from prison two years later, her uncle Georger engineered her betrothal to an old political ally, Alexander Pendarves, who happened also to have quarrelled with his nephew and heir, Francis Basset of Tehidy, because he refused on his inheritance to adopt the name Pendarves. That he was sixty years old, exceedingly unkempt and afflicted with gout did not matter. Mary would be sure of a fortune after his death.

Mary took one look at his 'large unwieldy person, his crimson countenance' and formed an instant aversion to him. She was well aware, however, of her father's financial straits and of his debt to her uncle, who had paid him an allowance for years, and she accepted her fate with as much patience as she could muster. But although Alexander was kind to her she could not overcome her dislike of him. The couple made a slow progress down to Cornwall, staying with friends on the way. After a tiring, uncomfortable journey made almost unbearable by the proximity of her husband, they arrived at Roscrow. At first sight of her new home, Mary 'fell into a violent passion of crying'. It must have appeared to her as the back of beyond.

This is how she described the old house:

> The castle is guarded with high walls that entirely hide it from your view. When the gate of the court was opened and we walked in, the front of the castle terrified me. It is built of ugly, coarse stone, old and mossy, and propt with two great stone buttresses, and so it had been for threescore years. I was led into an old hall that had scarce any light belonging to it; and on the lefthand of which was a parlour, the floor of which was rotten in places, and part of the ceiling broken down, and the windows were placed so high that my head did not come near the bottom of them.

p116   After the Restoration, Francis's son Sidney entered the royal household as a page and quickly won a reputation for honesty and financial acumen. He became Lord High Treasurer under Queen Anne and was created 1st Earl of Godolophin. Margaret Godolphin, his wife, died tragically young, and there is a moving account by the diarist John Evelyn of her last journey to Cornwall, to be buried at the family home. After describing the number of the entourage, and the various stages of the long passage by road, he concludes, 'The funeral can not have cost much less than £1,000.' Godolphin's political career came to an end in 1710 when he was dismissed after the Queen's quarrel with his friend the Duke of Marlborough, whose military campaigns he had financed. As Godolphin power and influence declined, so did the house. Sidney's only son Francis, 2nd Earl of Godolphin, married Henrietta Churchill, the Duke of Marlborough's eldest daughter, and they lived mostly at Newmarket since he was passionately fond of racing. The famous Godolphin Arab, shown in the painting by John Wootton which hangs in the dining-room, belonged to him.

Four of their five children died without issue, leaving Mary, the youngest, as successor. Mary married Thomas Osborne, the 4th Duke of Leeds, who changed his name to Godolphin-Osborne. Their heirs never lived in Cornwall.

p117   The occupations of Pengersick were perhaps not only defensive, to judge by the *St Anthony* affair. A ship of this name, a carrack of the King of Portugal, carrying a priceless cargo, was wrecked at Gunwalloe in 1526. The cargo vanished mysteriously as soon as the ship went aground, and became the subject of a Royal Commission. The first John Milliton's son, another John, together with William Godolphin and an unnamed neighbour, were implicated in the affair, but it seems no evidence was found to prove their involvement. John later became High Sheriff and Captain of St Michael's Mount - honourable positions for an honourable man. *The St Anthony*'s treasure was never recovered.

p121   Evidence exists of a flourishing trade between the Mount and the Continent. Diodorus, a Sicilian Greek historian writing early in the first century A.D., quotes extracts from a lost account of a voyage made round Spain and up to Britain in the fourth century B.C. by a Greek geographer called Pytheas, who was searching for the source of amber in the Baltic. 'The inhabitants of that part of Britain which is called Belerion [i.e. Land's End],' says Diodorus,

are fond of strangers and from their intercourse with foreign merchants are civilised in their manner of life. They prepare the tin, working very carefully the earth in which it is produced. . . They beat the metal into masses and carry it off to a certain island off Britain called Ictis. During the ebb of the tide the intervening space is left dry and they carry over to the island the tin in abundance in their wagons.

St Michael's Mount was probably the ancient island of Ictis (the only other, less likely, alternative being the Isle of Wight), and it was no doubt the Veneti, a powerful seafaring tribe from Brittany, who carried the tin over to Gaul in sturdy ships of oak with high prows and leather sails. Having reached the mouth of the Garonne (present-day Bordeaux) they travelled overland on horseback to the Mediterranean ports of Narbonne and Marseilles. Julius Caesar's destruction of the Veneti in a mighty naval battle off Brittany in 56 B.C. put an end to Cornish trading and cultural links with the Continent. The Romans found another source of tin in Spain, and for a period of almost eight hundred years Cornwall was a remote backwater largely unaware of the benefits Roman civilisation had brought to the rest of Britain and unaffected by subsequent Saxon plunder.

p122★★   The monastery never became a centre of learning and culture like its

French counterpart, partly because during the Hundred Years' War the Abbot and his monks, having sworn an oath of allegiance to Mont St Michel, were viewed with great suspicion and their revenues were often seized by the crown. Nor did the priory escape the ravages of the Black Death in 1349. By 1362 there were only two surviving monks and the prior in the monastery.

p122   Apart from its importance as a trading centre and place of pilgrimage, St Michael's Mount played its part in England's political history. In 1193, when Richard Coeur de Lion was imprisoned in Austria, having been captured by a fellow crusader on his way back from the third crusade, Henry de la Pomeray seized and occupied the Mount in the name of King Richard's brother John, Earl of Cornwall. Pomeray requested the unsuspecting monks to open the gates so that he could visit his sister in the monastery. Once inside, he and his men threw off their pilgrim's habits and forced the monks to give them possession of the various houses on the island. They lived there in style for some time, not thinking that King Richard would be ransomed for £100,000 and would return to London. At first Pomeray refused to surrender, but when an angry army arrived to besiege the Mount he gave himself up, dying, some say, of fright.

p122†   It was crucial for the Mount to be in loyal and trusted hands. As a military outpost at the most south-westerly tip of England it played a vital role in both attack and defence, when a hostile force approached by sea. In 1587 the beacon seen blazing from the church tower was the first of a chain of fires to warn of the approach of the mighty Spanish Armada.

Towards the end of her reign, Queen Elizabeth, in desperate need of funds to pay for the costly Spanish wars, sold the Mount to the Earl of Salisbury. In 1640 it was sold again to a staunch Royalist, Francis Basset, a portly but dependable cavalier to judge from the fine portrait of him by Cornelis Janssens which hangs in the Mount Museum today. The Bassets were an old and respected Cornish family who had lived at the manor of Tehidy, near Camborne, since the year 1200. Francis Basset strengthened the fortifications at his own expense and left his wife Anne in charge of the Mount when he went away to fight in the Civil War. Royalist ships that had managed to escape the guns of the Parliamentarian naval blockade brought ammunition from France into the friendly waters of Mount's Bay, such arms being paid for by the sale of Cornish tin. In 1644 the young Prince of Wales lodged at the Mount on his way to the Scilly Isles.

For a few years the Mount was safe, but Royalist fortunes turned in 1645. The Cornish troops, war-weary and embittered by their own squabbling commanders, looked in admiration at Cromwell's disciplined Model Army and lost their will to fight. Soon afterwards Parliamentary troops landed in a gap between the cliffs now known as Cromwell's passage. On April 23rd, 1646, Sir Arthur Basset, Francis's brother, surrendered. Relieved at being spared a long siege, the Parliamentary army allowed Basset and his officers to retreat to the Scilly Isles.

Some Royalist leaders chose exile rather than submission to Parliament; those who remained risked having their estates confiscated and were forced to pay huge fines. Many families were ruined and the impoverished

Bassets sold the Mount in 1659 to Colonel John St Aubyn, the Parliamentary leader who had been nominated Captain of the Mount twelve years earlier.

Alphabetical List of Places